THE MESSIER ALBUM

Messier's decorative colophon, inscribed by him in his
personal copy of his catalogue.

THE MESSIER ALBUM

By

JOHN H. MALLAS

and

EVERED KREIMER

FIRST EDITION

Sky Publishing Corporation

Cambridge, Massachusetts

Fourth printing, 1987

Library of Congress catalog card number 78-63243

Library of Congress Cataloging in Publication Data

Mallas, John H
   The Messier album.

   Includes a facsim. reproduction of Messier's Catalogue des nébuleuses et des amas d'étoiles (p. 18)
   Bibliography: p. 216
   1. Nebulae — Observations. 2. Stars — Clusters — Observations. 3. Messier, Charles. I. Kreimer, Evered, joint author. II. Messier, Charles. Catalogue des nébuleuses et des amas d'étoiles. 1980. III. Title.
QB851.M34 1980   523.8'9   80-17957
ISBN 0-933346-04-2

Printed by Murray Printing Co., Westford, Massachusetts

Contents

The famous compiler of the first extensive catalogue of nebulae and star clusters, Charles Messier (June 26, 1730-April 12, 1817), as painted by Desportes in March, 1771. The astronomer is shown at the age of 40, about the time his first memoir and the catalogue were presented. On the back of the portrait Messier had written, "This portrait is a good likeness, except that I appear younger than I am, and I have been given a better expression than I have."

Messier and His Catalogue

OWEN GINGERICH

Harvard-Smithsonian Center for Astrophysics

IN THE MIDDLE of the 18th century, astronomy stood at the threshold of a great observational age. Flamsteed and Bradley had already made their important contributions to positional astronomy. Halley, Roemer, and Hevelius were names of history. Yet the real telescopic observers were still to come. In 1758 Dollond would perfect the achromat, and within a few decades the great reflectors of Herschel would probe into the unknown depths of the heavens. But it would be for a Frenchman, Charles Messier, whose first love was comets, to build the foundation for telescopic sidereal astronomy with his catalogue of nebulae and star clusters.

Messier, with his quaint mixture of the old and new, typifies the enthusiasm with which astronomers met the improved techniques. To 20th-century astronomers his fascination for comets appears misguided; with modern concepts of the vastness of the universe, comets appear relatively insignificant. But to the astronomer of 200 years ago, it was the static, nondescript nebulae which were trivial. To the 1750 astronomer, only about 50 comets were known since the beginning of time; within the next half century Messier personally observed more than that number.

The successful recovery of Halley's comet in 1759, a great triumph for the Newtonian theory, added further impetus to the study of comets. Messier's first fame, in fact, came in connection with this comet. Almost eight years earlier, in October, 1751, as an orphan 21 years old, he had come to Paris from his home at Badouvillier, Lorraine, to seek his fortune. The astronomer Joseph Nicolas Delisle hired him as a draftsman, and as a recorder for astronomical observations. Messier first set to work copying a map of the great wall of China. For this, which required a large space, Delisle assigned him a long hall in the College of France. The unheated corridor was rather uninviting in winter, but, as Delambre later commented, this was appropriate for a future observational astronomer. For, in addition to the drafting work, Messier was instructed in the use of the astronomical instruments. By 1754 he was an experienced ob-

1

server, handling the majority of the work. Shortly afterward, Messier obtained the position of clerk at the Marine Observatory in Paris.

At this time astronomers were anticipating the first predicted return of Halley's comet. Delisle had a large map drawn, showing the routes by which the comet would have to travel to arrive at perihelion on various days. No one doubted that by this scheme Delisle's observing assistant, Messier, would be the first to discover the comet, and then the pair would hasten to announce it.

For 18 months Messier systematically searched, but without success, for actually he had been misled by Delisle's careful maps! Thus, on Christmas night, 1758, a gentleman farmer in Saxony named Palitzsch made the first observation of the returning object. Some days later it was observed by others in Germany as well. But by an extraordinary chance, the news of the discovery was delayed over three months in reaching France.

Meanwhile, a month after the original discovery, on January 21, 1759, Messier independently found the comet. Delisle, who had become a gouty old individualist, refused to permit his assistant to announce the observations for two months, until the news of Palitzsch's discovery became known in France. By then the comet was lost in the twilight at perihelion passage, and the other astronomers, enraged at Delisle's obstinacy, refused to consider as authentic Messier's observations up to that time. On the other hand, Delisle would not believe the story of the "poor Saxon peasant who accidentally found the comet with his naked eye." The story is occasionally repeated today, although actually Palitzsch had used a telescope and was specifically searching for the comet. Delisle followed the same reproachable secretive procedure with two other Messier comets; his conduct is all the more inexplicable as he did not use the observations for any particular purpose.

Shortly afterward Delisle retired; Messier continued his work from the tower observatory at the Hotel de Cluny in Paris. He discovered the comet of 1764, and by a chance naked-eye view he saw that of 1766. For about 15 years nearly all comet discoveries were made by Messier, so that he almost considered them his own property. One perhaps legendary anecdote concerns the death of his wife, which prevented the discovery of a comet he had been watching for. Messier was in despair when his rival, Montagne de Limoges, found the object first. When someone spoke to him of the loss he had suffered, Messier would reply, always thinking of the comet, "Alas! I have discovered a dozen of them; Montagne had to take away the 13th!" And then tears would come to his eyes; remembering he should be weeping for his wife, he would exclaim, "Ah, the poor woman!" Altogether Messier claimed to have discovered 21 comets in his lifetime; modern astronomers with their more discriminating

The Hotel de Cluny as it appeared in Messier's day. It was built in the late 15th century by Abbot Jacques of Amboise, as a temporary residence for abbots of the Cluny order and their guests. In 1748, Delisle became one of the residents, and from his death in 1768 to 1817 Messier had his residence as well as observatory in the building. Lalande also lived here part of this time. The Marine Observatory was in the solid octagonal tower. From the *Bulletin* of the French Astronomical Society, 1917.

standards of what constitutes a discovery would reduce this number to perhaps 15.

But comets alone did not claim Messier's time. He was an ardent observer of occultations, transits, and eclipses, and his notebooks were filled with sunspot and meteorological observations as well. Yet seldom, if ever, did Messier do more than tabulate the observations. For instance, invariably it was his associates who reduced his comet positions, to determine the orbital elements.

Nevertheless, Messier's comet discoveries brought him considerable fame. He aspired to become a member of the exclusive Académie Royale des Sciences, but its scholars were reluctant to admit a mere observer. On several occasions Messier's name was proposed, even for other sections than astronomy. In 1763 Messier tied with Jeaurat for an open place in the academy, but on the second ballot lost. Meanwhile his fame spread outside France, so that after the death of Lacaille in 1762 he was generally considered the leading French astronomer. In 1758 he was already a member of the Royal Society in London, and later a map of a comet sent to the King of Prussia brought him membership in the Berlin academy. At the recommendation of La Harpe in Russia, Messier was named to the academy of St. Petersburg. Within the next few years he joined half a dozen other societies. By this time the French academicians took more notice of Messier, and he was finally admitted, in 1770.

THE ORIGINAL MESSIER CATALOGUE

Almost immediately Messier contributed the first of a great number of astronomical memoirs to the journal of the academy. The very first of these provides his most enduring bid to fame, for this was the "Catalogue des Nébuleuses et des amas d'Etoiles, que l'on découvre parmi les Etoiles fixes, sur l'horizon de Paris," the first installment of the famous Messier catalogue.

The list of nebulae had been a long time in the making. The first object, the Crab nebula in Taurus, was discovered on August 28, 1758, while Messier was following a comet he had found two weeks earlier. In his memoir he wrote, "When the comet of 1758 was between the horns of Taurus, I discovered above the southern horn and a short distance from the star Zeta Tauri a whitish light, extended in the form of the light of a candle, and which contained no stars. This light was a little like that of a comet I had observed before; however, it was a little too bright, too white, and too elongated to be a comet, which had always appeared to me before almost round, without the appearance either of a tail or a barb." As was the case for many of the later objects, the nebula in Taurus was duly plotted on the appropriate comet chart.

The next object in the list, the globular in Aquarius, was plotted on the chart of Halley's comet, although it was not actually observed until 1760. The task of compiling such a catalogue did not begin in earnest until 1764, however. Within seven months Messier sought out 38 more entries for the list. Included were such objects as the globular cluster in Hercules (M13), the Omega and Trifid nebulae in Sagittarius (M17 and M20), the Dumbbell planetary nebula (M27), and the Andromeda galaxy (M31).

At the end of this period, in October, he made a thorough search for previously reported nebulae in order to make his list as complete as possible. The chief lists that had been published before were a brief catalogue of five objects by Edmond Halley; a longer tabulation by William Derham of Winchester, England, which was primarily gleaned from Hevelius' star catalogue (*Prodromus Astronomiae*), and available to Messier in a French translation published by P. de Maupertuis in the Paris *Mémoires* for 1734; and the list of southern nebulae by N. de Lacaille in the *Mémoires* for 1755.

Messier located some of the objects in these compilations, but not all. He therefore published as part of his memoir a number of objects previously reported, but which he could not find. He realized that many of them were nebulous to the naked eye, but were mere asterisms when examined with a telescope. One of these, near the star 70 Ursae Majoris, he listed as No. 40, but this final listing from his 1764 efforts is frequently omitted from present-day editions of the Messier catalogue.

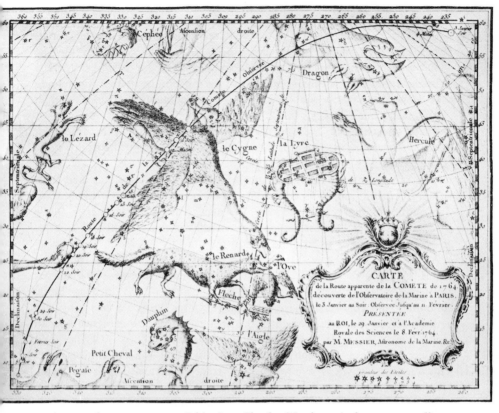

Among the astronomers of his day, Charles Messier was famous as a dis-
coverer of comets, including that of 1764. The paths of comets were
recorded on elaborate, specially drawn charts; this one by Messier was
published in the same book as his original catalogue, the 1771 volume
of the memoirs of the French Royal Academy of Sciences.

In January of 1765, Messier by chance found the galactic cluster near
Sirius (M41), but otherwise he made no systematic search, apparently.
On March 4, 1769, he determined the positions of three well-known objects: the Orion nebula (M42 *and* M43), the Pleiades, and Praesepe, evi-
dently adding these as "frosting" to bring the list to 45. Although
Messier had independently discovered the majority of the objects, a large
number, including the Crab nebula, had been found previously, a fact
generally recorded by him in each instance. Nevertheless, this first list
neglected several southern objects reported in 1755 by Lacaille and included in the extension of the Messier catalogue published several years
later.

The initial list of 45 objects appeared in the *Mémoires de l'Académie* for 1771, which was actually published in 1774. In the same year the catalogue was also printed in Lalande's *Ephémérides* for 1775-84. This first group of nebulae was the only part of the catalogue prepared for its own sake; most of the subsequent objects were found and reported in comet searches, so that the remainder of the catalogue is less systematic than the first part. There is no significant difference in limiting magnitude for the objects in the first group compared with the later observations, for Messier's instruments remained about the same.

These telescopes of Messier, a source of great curiosity to amateurs, make a puzzling historical problem, for in the memoir he follows the standard practice of the day in listing his instruments by length and magnifying power, instead of by aperture. The reflecting telescopes had inefficient speculum mirrors, adding further uncertainty to the amount of light-gathering power. Fortunately, however, Messier's friend J. S. Bailly used a number of these telescopes for a study of the satellites of Jupiter, and Bailly has given us invaluable information.

At one time or another, Messier used over a dozen telescopes, but none larger than his favorite, a 104-power Gregorian reflector. This instrument had a length of 32 inches and an aperture of 7½ inches. Bailly computed that this was equivalent to a refractor about 28 feet long and with an effective aperture of about 3½ inches! Even less effective was an old 8-inch octagonal Newtonian, which had belonged to Delisle, and which was undoubtedly the original instrument at the Hotel de Cluny. This was rarely used, for it was less efficient than a 2½-inch refractor. Messier frequently mentioned small simple refractors. Later he observed with several 3½-inch achromats, for at this time achromatic refractors were just coming into use. Those of Messier generally were 3½ feet long, with a magnifying power of about 120.

Messier voluntarily chose the smaller instruments, whereas Herschel, only a little later, exploited large reflectors. Messier's choice naturally arose from his interest in comets; Herschel, however, was able to resolve over a score of clusters which appeared merely nebulous to Messier. As an admirer of Messier's list, Herschel carefully avoided giving his own numbers to any object catalogued by the Frenchman.

THE FIRST SUPPLEMENT

On February 19, 1771, three nights after Messier had presented his first memoir to the academy, he recorded the positions of four more clusters. These were the first of 23 nebulae that were to be published in 1780 as a supplement to the original list.

Within the next few years, half a dozen more objects were found, mostly while he was observing comets, and these were indicated on the beautifully engraved charts accompanying the comet memoirs. The discovery of an object in Coma Berenices was specifically chronicled before the others, in connection with a memoir on an unusual aurora borealis in 1777. In the catalogue, however, it appears in order of discovery as M53. Several months later Messier finally located an object in Sagittarius previously reported by Lacaille, but which Messier had been unable to find in 1764, and which was then reported in his list of nebulae not found. This globular cluster became the 55th entry.

Beginning in 1779, Messier observed nine more objects in connection with the comet of that year. This comet passed across the Coma-Virgo region of galaxies, bringing about the first discoveries in that area. In the following year, by chance he found the 65th and 66th objects, a pair of galaxies in Leo. In 1773 a comet had passed between them, but Messier failed to note their presence because of the light of the comet.

In April of 1780, Messier made the final two observations to bring the list up to 68 in time for the deadline of the French almanac, *Connaissance des Temps*. Although this almanac consisted primarily of standard tables, a few original contributions, mostly of numerical material, were incorporated in each volume. Thus Messier's 1771 catalogue of 45 nebulae and clusters, cast in tabular form, was included in the almanac for the year 1783, along with the 23 new discoveries. The objects were listed with positions, and the date when each position was determined, and on the facing page a brief description was given. The editors noted, "In addition to the catalogue published by Messier, which we give here, we report further a large number of nebulae and clusters which he has discovered since the printing of the memoir, and which he has communicated to us."

After the catalogue proper came the list of previously mentioned nebulae that Messier was unable to locate, followed by the brief tabulation of the 42 southern nebulae observed by Lacaille at the Cape of Good Hope.

Of the first three objects recorded by Messier in the supplement to his original list, two of them, M47 and M48, have been a mystery to succeeding astronomers. The descriptions of M46-48, translated from the *Connaissance des Temps* for 1783, are:

46. 7h 31m 11s, —14° 19' 07". Cluster of very faint stars, between the head of the Great Dog and the two hind feet of the Unicorn, located by comparing this cluster with the 6th-magnitude star 2 Navis (according to Flamsteed); these stars can be seen only with a good refractor; the cluster contains a little nebulosity.

47. 7h 44m 16s, —14° 50' 08". Cluster of stars a short distance from the preceding; the stars are brighter; the middle of the cluster was compared with the same star, 2 Navis. The cluster contains no nebulosity.

48. 8h 02m 24s, —1° 16' 42". Cluster of very faint stars, without nebulosity; this cluster is a short distance from the three stars that form the beginning of the Unicorn's tail.

The position given for M47 falls in a comparatively empty part of a rich Milky Way region. By successive copying, this spurious position received the number NGC 2478, although no astronomer since Messier was able to confirm the observation. There is, however, a fine cluster in this region, brighter than M46 and so close to it that Messier could hardly have overlooked it. This cluster, NGC 2422, was identified as M47 by Oswald Thomas in his *Astronomie* (1934).

More explicit reasons for this identification were given independently in 1959 by T. F. Morris, a member of the Messier Club of the Royal Astronomical Society of Canada's Montreal Centre. Dr. Morris suggested that an error in signs in the differences between M47 and the comparison star could account for the position.

Messier determined the declination of a nebula or cluster by measuring the difference between the object and a comparison star of known declination. The right ascension could be found by recording the times at which the object and the star drifted across a central wire in his telescope's field; the time interval gives the difference in right ascension. The differences between Messier's 1770 position for M47 and his stated comparison star, 2 Navis (now 2 Puppis), if applied with opposite signs, lead to NGC 2422. Clearly, Messier made a mistake in computation!

Although the circumstances of M48 are not as obvious, only one cluster of the size and brightness likely to be recorded by Messier is found in the region near "the three stars that form the beginning of the Unicorn's tail" (Zeta, 27, and 28 Monocerotis). Dr. Morris has pointed out that this cluster, NGC 2548, has the same right ascension as the position given for M48. (Allowance must, of course, be made for precession in comparing Messier's figures with modern positions.)

The declinations disagree by about five degrees. Since no conspicuous star is located 2½ degrees away in declination, we cannot account for this position by another error in sign. It seems unlikely that the comparison star was misidentified, since the right ascension is probably correct. Messier did not publish the name of the star used, and his original records are apparently no longer extant. I have examined all of his observing books preserved at the Paris Observatory, but none are included from this period.

Thus, a careful survey of the region described by Messier leads to the conclusion that NGC 2548 is the cluster that the French observer intended as his 48th object, for lack of any other cluster nearby that fits the description.

MECHAIN AND THE SECOND SUPPLEMENT

More and more frequently in the 1780's a new name appeared in the annals of comet discoveries. While Messier's indefatigable observations had earned him the title of "ferret of comets" from Louis XV, he found keen competition in a younger astronomer at the Marine Observatory, Pierre Francois André Méchain. It would not be surprising to discover a strong jealousy between these rivals. Yet, if it existed, biographers have remained silent. We can only deduce that the two were professional acquaintances, freely sharing their data. Several times Messier recorded that "Méchain informed me of the discovery [of the comet] the following day." Their careers have so many points of contact that any biography of Messier must also be an account of Méchain.

Born August 16, 1744, in Laon, Méchain was slightly over 14 years Messier's junior. As a boy, he studied mathematics and physics as well as architecture, his father's profession. Méchain later went to Paris, to the leading architectural school. But financial difficulties beset his father, and to raise funds Méchain's astronomical quadrant was offered to the astronomer Lalande, who immediately became interested in the young man who had purchased it in his desire to study the heavens.

After an interview, Lalande agreed to teach Méchain astronomy, and assigned him the proof sheets of the new edition of his *Astronomie.* When Lalande visited him shortly thereafter, he was astonished at Méchain's progress in the most difficult astronomical mathematics. In 1772 Lalande procured Méchain a position of astronomer-hydrographer in the naval map archives (Dépot de la Marine) at Versailles. But the position was uncertain, because of the political unrest in the French government, and Lalande obtained for him employment as a tutor. Eighteen months later the entire department was reorganized, and Méchain returned to his position. Soon thereafter the Dépot was moved to Paris, where, if not before,

Messier's contemporary, Pierre Méchain (1744-1805), from a portrait in the *Monatliche Correspondenz,* 1800, edited by Baron F. X. von Zach.

9

Méchain must have met Messier, who was well established as Marine Astronomer. Méchain's work called him away from Paris much of the time, however, as he was mapping the French coastline.

Meanwhile Méchain had turned his eyes to the heavens, and in 1774, while still at Versailles, he observed an occultation of Aldebaran. Lalande introduced Méchain to the Academy of Sciences by presenting a report of the observation, which was later printed in the memoirs. In 1781, Méchain discovered two new comets; in addition to observing them, he calculated their orbits. The following year the academy awarded Méchain a prize for his essay on the comets of 1552 and 1661. A return was expected in 1789, but he showed that these were two separate comets, neither of which would return that year. Such theoretical work was typical of Méchain, and calculations as well as observations were generally included in his comet memoirs. Thus the mere discovery of about 10 comets, including Encke's and Tuttle's, represents only a small part of his efforts. Méchain was soon recognized for his work, and in 1782 he became a member of the academy.

During his comet searches in 1780-81, Méchain discovered about 32 new nebulous objects. Appropriately, these discoveries were communicated directly to Messier, who checked their positions in the course of his own observing, but not always in their original order, so for a further extension of the catalogue each one was numbered as Messier observed it. He generally followed up Méchain's observations several weeks later, but as the deadline for the 1784 *Connaissance des Temps* approached he confirmed them within a few days. Apparently Méchain first realized the abundance of nebulae in the Coma-Virgo region. On March 18, 1781, Messier followed up the lead with a busy night, recording nine new nebulae in the area.

Of Messier's 14 nebulae in this region, 13 are easily identified today, but M91 cannot be found. Is it NGC 4571, as the *New General Catalogue* suggests? Photographs indicate that this galaxy is much too faint for Messier's small telescopes. Inspection of Palomar Sky Survey photographs shows that Messier was astonishingly successful in locating all the bright galaxies in this area, while he invariably missed the fainter ones. There is simply no bright nebula omitted by Messier that could conceivably be identified with M91.

We look in vain at Messier's description of his 91st object for further clues:

89. 12h 24m 38s, +13° 46' 49". Nebula without stars, in Virgo, a short distance from and on the parallel of declination of the nebula reported above, No. 87. Its light is extremely weak and faint, and it is not possible to detect it without difficulty.

90. 12h 25m 48s, +14° 22' 50". Nebula without stars, in Virgo; its light is as faint as the preceding, No. 89.

91. 12h 26m 28s, +14° 57' 06". Nebula without stars, in Virgo, above the preceding, No. 90; its light is still fainter than that above.

Could this object have been a comet, as Harlow Shapley and Helen Davis suggested? Possibly, but I believe that another explanation is more likely. In each of Messier's three groups of nebulae recorded in the Coma-Virgo region, the listing proceeds in order of right ascension. One can imagine Messier using a technique frequently employed by amateur observers today. Quite probably he scanned the sky along a north-south line, letting the diurnal motion of the heavens slowly carry the more easterly stars into his telescope's field of view for the next sweep. In this way he could detect the nebulae one after the other, and by timing their transits across a central wire, he could compute their right ascensions.

If this were the case, M91 would be properly identified with the next nebula east of M90 to be caught in the north-south sweeps in declination. Not surprisingly, there is a bright nebula with a right ascension almost exactly the same as M91's — the nebula Messier had recorded two years earlier as M58. Furthermore, the difference in declination is about two degrees. The comparison star used for these nebulae is not known, but presumably Messier determined the declinations by differences, rather than directly from a circle. These objects are so faint and diffuse that Messier must have experienced difficulty in getting precise positions. Considering this uncertainty, it is possible that the error in declination could have been *exactly* two degrees, and thus be accountable as a simple mistake in the reduction of the observations.

Although the evidence is circumstantial, I believe we can reasonably assume that the listing of M91 has resulted from a duplicate observation of M58.

Messier made his last catalogue observation on April 13, 1781, only a year after his final work on the 1780 addition to his original list. This brought the total number to 100, with 24 of these additions credited to Méchain. Unfortunately, Messier had no chance to check three of Méchain's observations, or positions for two more objects in Ursa Major noted with the description of M97 (see the quotation later in this chapter), before publication of this list (in 1781). Nevertheless, he attached Méchain's observations to the end as 101, 102, and 103, although no positions were given for the latter two. Thus, 35 new objects and descriptions were added to the previous 68, and the earlier descriptions were slightly extended to cover more recent observations.

Further revision of the catalogue was abruptly terminated by a severe accident to Messier. In November, 1781, he visited the garden of

Monceau with his friend Bochard de Saron, president of the French Assembly. Messier's attention was attracted by a door which he assumed led to a grotto. Instead, it was an icehouse. Entering without precaution, he fell 25 feet onto the ice, breaking his arm, thigh, wrist, and two ribs. Fortunately, someone noticed with surprise that the icehouse door was open. Three men with ropes and ladders finally removed him, and the astronomer was placed under the attention of the most skilled academy surgeons. Messier, always an empirical observer, disliked the theory and art of the surgeons, especially when Valdajou rebroke his thigh to set it better!

President Saron and the other academicians were especially solicitous during Messier's illness. A Monsieur Sage obtained a royal grant of 1,200 livres for him, followed by a pension of 1,000 livres and another gift of 2,400 livres. President Saron, incidentally, was well known for his exemplary generosity in science, and was an accomplished mathematician. He was among the first to realize, by computing its orbit, that Uranus was a new planet.

MECHAIN'S SUPPLEMENT

One year and three days after his fall, Messier again entered his observatory, to prepare for a transit of Mercury. Soon he resumed his observations of Herschel's new planet that had been interrupted by the accident. His tireless watch of comets, eclipses, and transits began again.

Further work to extend the catalogue of nebulae and clusters was also planned, although in 1784 the previous list was printed exactly (in the *Connaissance des Temps* for 1787), including most of the same typographical errors. As before, no positions were given for M102 and M103, and hence later observers sometimes did not include them in Messier catalogue lists. In 1783, for instance, the German almanac, Bode's *Berliner Astronomisches Jahrbuch* for 1786, published a list of the objects from M46 to M100 with positions determined by Messier, and in Bode's *Uranographia* (1801) only the initial 100 appeared.

The description given for M102 was insufficient to locate the nebula unambiguously, and considerable confusion has resulted from attempted identifications. Yet this mystery can be solved with complete certainty. Soon after the publication of the second supplement to Messier's catalogue, which included M102, Méchain sent a correction to J. Bernoulli in Berlin, explaining that M102 was identical to M101. The confusion had resulted from an error in his star chart. Méchain's letter was translated into German and printed in the *Berliner Astronomisches Jahrbuch* for 1786, the same volume that contained the copy of the latter part of Messier's catalogue. The real puzzle is why this letter was overlooked by so many astronomers for so long.

12

In 1877, E. Holden included a reference to Méchain's letter in his useful *Index Catalogue of Books and Memoirs Relating to Nebulae and Clusters*. The correction is mentioned even more explicitly by G. Bigourdan in his thorough analysis of old discoveries of nebulae in *Observations, 1907*, of the *Annales* of the Paris Observatory (published in 1917). Nevertheless, other investigators of Messier's catalogue seemed unaware of the correction until 1947, when Helen S. Hogg published Méchain's letter in the *Journal* of the Royal Astronomical Society of Canada. A slightly abbreviated version of the French text is found in the *Histoire* of the Berlin academy for 1782 and has been overlooked even longer than its German translation.

Along with the correction concerning M102, Méchain also included the descriptions of six new objects. Four of these had been found too late for inclusion in the 1784 *Connaissance des Temps*. Positions were given for three of them, while the descriptions suffice to locate the others.

Undoubtedly Messier was aware of the new discoveries almost as soon as they were made. Evidence is found in his personal copy of the printed list. A handwritten note at the bottom of a page indicates another nebula in Virgo, one of the four new ones reported by Méchain in the *Jahrbuch*. Several other personal notations are of interest. The printed description of M97 (the Owl nebula) reads in part, "near by this nebula is seen another, which has not been measured, and also a third which is near Gamma of Ursa Major"; following this, Messier has written in the position of NGC 3992. Oddly, positions have been added by hand for both M102 and M103, in spite of Méchain's claim that 101 and 102 were identical. (There is no obvious object at this erroneous position for M102.) Some of the notations were made in the spring of 1790, proving that this copy was still in use many years after the error was known.

The nature of the added remarks in Messier's copy indicates that they were in part additions made soon after the original publication, and were in part intended for a revision to be published after 1790. Political turmoil in France and inflation prevented publication during that time. It seems odd, however, that a revised list had not been published during the six years (1785-1790) immediately following republication of the list in the 1787 almanac, when Méchain himself served as editor.

The French revolution brought trying times to Messier, as well as to his colleagues, many of whom were scattered from Paris. Soon after he had become a pensioned academician, the Académie Royale was suppressed. Messier lost his pensions and his salary from the navy. At the same time, the navy stopped paying the rent of his observatory at the Hotel de Cluny. Fortunately, he had some savings, but his means were very limited. Messier was forced to go to Lalande to get oil for the lamp he used during his observing. In September of 1793 he discovered another comet, sending his observations to Saron who finished computing the orbit in prison a few days before he was guillotined.

When, in 1791, the National Assembly decided to introduce a general metric system of measurement, Méchain was chosen to survey the southern half of the base line from Dunkerque to Barcelona from which the standard meter was derived. For seven years he worked in Spain and the Pyrenees. Although he was away from Paris during the bloodshed there, his family remained in that city and he lost all savings and his estate.

Fortunes improved for both astronomers, however. For a few years before his death in 1805, Méchain was director of the Paris Observatory. Both Méchain and Messier became members of the new Academy of Sciences, and the Bureau of Longitudes. From Napoleon, Messier received the cross of the Legion of Honor. He had no children, but during his last 19 years his niece gave him assiduous care. After a stroke his strength diminished, and he spent the last two years at home. He lived to the age of 86, dying April 12, 1817.

Looking back on his interest in nebulae, Messier wrote in the *Connaissance des Temps* for 1801: "What caused me to undertake the catalogue was the nebula I discovered above the southern horn of Taurus on September 12, 1758, while observing the comet of that year. . . . This nebula had such a resemblance to a comet, in its form and brightness, that I endeavored to find others, so that astronomers would not confuse these same nebulae with comets just beginning to shine. I observed further with the proper refractors for the search of comets, and this is the purpose I had in forming the catalogue. After me, the celebrated Herschel published a catalogue of 2,000 which he has observed. This unveiling of the sky, made with instruments of great aperture, does not help in a perusal of the sky for faint comets. Thus my object is different from his, as I only need nebulae visible in a telescope of two feet [length]. Since the publication of my catalogue I have observed still others; I will publish them in the future, according to the order of right ascension, for the purpose of making them more easy to recognize, and for those searching for comets to remain in less uncertainty."

What were the other, newly observed objects? A handwritten list of the catalogue objects in order of right ascension is included in his personal copy of the catalogue, but with no additions. Undoubtedly the objects reported by Méchain were among the new ones, although Messier never mentioned these in print. In fact, Messier himself, to my knowledge, never published further information about new nebulae and clusters, with a single exception. One of his last memoirs, in 1807, discussed the Andromeda galaxy and two faint companions. A fine engraving illustrated all three nebulae. Only one of the companions, M32, was included in

14

the original list, although the second (NGC 205) had been found by him as early as August 10, 1773.

In 1921, when Camille Flammarion worked intensively on the Messier catalogue, he found an addition in Messier's personal working copy, and designated this nebula in Virgo (NGC 4594) as M104. After Dr. Helen Sawyer Hogg brought to light Méchain's letter to the *Berliner Astronomisches Jahrbuch,* she identified four of Méchain's new objects. Since one was the same as Flammarion had found, she suggested that the other three should be numbered consecutively M105 to M107.

Although Flammarion found Messier's notation of the position of the nebula near Gamma Ursae Majoris, mentioned earlier in this chapter, he made no attempt to number it, and because Méchain did not give precise positions, Dr. Hogg omitted identifications of this and the other nebula near Beta Ursae Majoris. From my study of this region, the nebula near Beta is unambiguously NGC 3556, while an examination of the critical limiting magnitude of the catalogue indicates that the one near Gamma must be NGC 3992, a fact confirmed by the position Messier added to his personal copy. Thus, if the objects from M104 to M107 are included, it seems logical to me to number NGC 3556 and NGC 3992 as M108 and M109 respectively, especially since they are mentioned in the original catalogue.

But we have only circumstantial evidence that Messier ever observed any of these, and none of them received his sanction in print, nor did Bode include these new positions with the Messier list when he printed Méchain's letter. Furthermore, we have evidence that Messier knew of other nebulae, specifically NGC 205 in Andromeda, which recently has been proposed as M110. Finally, contemporary workers, such as the Herschels, never considered as part of the Messier list any objects beyond the final 103 in the *Connaissance des Temps* for 1784 and 1787. If the asterism M40 and the duplications M91 and M102 are excluded, then, in my opinion, the Messier catalogue contains exactly 100 objects. The nebulae and clusters M104-109 should then be referred to as "Mechain 104-109." What to do about NGC 205 is still moot.

Why did Messier never publish his own revised list? Perhaps his perfectionist nature prevented the release of an incomplete list. More likely, he probably never got around to it. Although his observations and reductions continued, he was not always well and his sight began to fail in later years. In 1802, when Messier was 72, Herschel visited Paris, and wrote in his diary, "A few days ago I saw Mr. Messier at his lodgings. He complained of having suffered much from his accident of falling into an ice cellar. He is still very assiduous in observing, and regretted that he had not interest enough to get the windows mended in a kind of tower where his instruments are; but keeps up his spirits. He appeared to be a

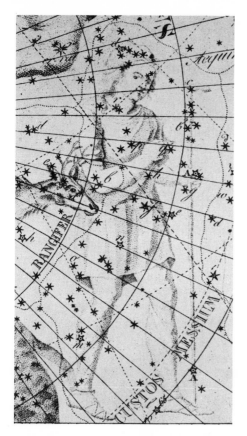

very sensible man in conversation. Merit is not always rewarded as it ought to be."

In 1775 Lalande formed a constellation Messier on his new globe. Later it was included in the Paris edition of Flamsteed's atlas, and Bode's *Uranographia,* reproduced in part here. Wrote Lalande, "A 'messier' [keeper of the crops] is one who superintends the guarding of the grain. The constellation which carries this name is between Cepheus, Cassiopeia, and Camelopardalis, that is to say, between the rulers of an agrarian people and an animal destructive to the crops."

The constellation Messier in Bode's atlas of 1801.

REFERENCES AND ACKNOWLEDGEMENTS

Extensive information about Charles Messier can be found in Delambre's *Histoire de l'Astronomie au Dix-huitième siècle* (Paris, 1827) and in Lalande's *Bibliographie Astronomique* (Paris, 1803). Messier's own detailed accounts of his observations appear in the volumes of *Connaissance des Temps* for 1799, 1800, 1801, 1807, 1809, and 1810. A list of Messier's publications appears in Quérard's *La France Litteraire* (Paris, 1827-1839). A discussion of the entire catalogue, including Messier's original descriptions, was published as a series by Camille Flammarion in the *Bulletin* of the Astronomical Society of France from November, 1917, to August, 1921.

I wish to thank Dr. Helen Sawyer Hogg and Miss Sally Hogg for several major translations used for this study, Prof. André Danjon for arranging for me to examine the Messier manuscripts at the Paris Observatory, the permanent secretaries of the French Academy of Sciences for permission to inspect Messier's dossier in the academy archives in July, 1955, and Bernard Roy for the picture of Messier.

O. G. [Originally published in 1953 and 1960]

Messier's Own Catalogue

R EPRODUCED IN FACSIMILE on the following pages is the Messier catalogue as Messier himself published it in the *Connaissance des Temps*.

In this historic version, reduced to two-thirds the original size, each left-hand page of tabular matter gives, in successive columns: the date of observation; the Messier number of the object; right ascension in time and also in degrees, minutes, and seconds; declination, marked *A* if south and *B* if north; and the diameter of the object in minutes and seconds of arc. The coordinates are in each case for the date of observation, instead of being reduced to the same equinox. For the objects first seen by Méchain, his positions are also given by Messier.

The right-hand pages contain Messier's own descriptions of his objects, as seen in the small telescopes he used. Some frequently occurring phrases are: *amas d'étoiles,* cluster of stars; *amas de petites étoiles,* cluster of faint stars; *nébuleuse sans étoile,* starless nebula. Whenever a telescope is mentioned, Messier gives its length rather than its aperture, as for example in connection with M5, where *on la voit trés-bien, par un beau ciel, avec une lunette ordinaire d'un pied* means "It is seen very well in a good sky with a nonachromatic refractor one foot long."

DATE des OBSERVATIONS.	Numéros des Nébuleuses	ASCENSION DROITE.		DÉCLINAISON.	Diamètre en degrés & min.
		En Temps.	En Degrés.		
		H. M. S.	D. M. S.	D. M. S.	D. M
1758.Sept.12	1.	5.20.2	80.0.33	21.45.27 B	
1760.Sept.11	2.	21.21.8	320.17.0	1.47.0 A	0.4
1764. Mai 3	3.	13.31.25	202.51.19	29.32.57 B	0.3
8	4.	16.9.8	242.16.56	25.55.40 A	0.2½
23	5.	15.6.36	226.39.4	2.57.16 B	0.3

Détails des Nébuleuses & des amas d'Étoiles.
Les positions sont rapportées ci-contre.

1. Nébuleuse au-dessus de la corne méridionale du Taureau, ne contient aucune étoile; c'est une lumière blanchâtre, alongée en forme de la lumière d'une bougie, découverte en observant la Comète de 1758. Voyez la Carte de cette Comète. *Mém. Acad. année 1 59, page 188*; observée par le Docteur Bévis vers 1731. Elle est rapportée sur l'*Atlas céleste* anglois.

2. Nébuleuse sans étoile dans la tête du Verseau, le centre en est brillant, & la lumière qui l'environne est ronde; elle ressemble à la belle Nébuleuse qui se trouve entre la tête & l'arc du Sagittaire, elle se voit très-bien avec une lunette de deux pieds, placée sur le parallèle de α du Verseau. M. Messier a rapporté cette nébuleuse sur la Carte de la route de la Comète observée en 1759. *Mém. Acad. année 1760, page 464.* M. Maraldi avoit vu cette nébuleuse en 1746, en observant la Comète qui parut cette année.

3. Nébuleuse découverte entre le Bouvier & un des Chiens de Chasse d'Hévélius; elle ne contient aucune étoile, le centre en est brillant & sa lumière se perd insensiblement, elle est ronde; par un beau ciel on peut la voir avec une lunette d'un pied; elle est rapportée sur la Carte de la Comète observée en 1779. *Mémoires de l'Académie de la même année.* Revue le 29 Mars 1781, toujours très-belle.

4. Amas d'étoiles très-petites; avec une foible lunette on le voit sous la forme d'une nébuleuse; cet amas d'étoiles est placé près d'*Antarès* & sur son parallèle. Observé par M. de la Caille, & rapporté dans son Catalogue. Revu le 30 Janvier & le 22 Mars 1781.

5. Belle Nébuleuse découverte entre la Balance & le Serpent, près de l'étoile du Serpent, de sixième grandeur, la cinquième (suivant le Catalogue de Flamstéed); elle ne contient aucune étoile; elle est ronde, & on la voit

DATE des OBSERVATIONS.	Numéros des Nébuleuses	ASCENSION DROITE.		DÉCLINAISON.	Diamètre en degrés & min.
		En Temps.	En Degrés.		
		H. M. S.	D. M. S.	D. M. S.	D. M
1764. Mai 23	6.	17.24.42	261.10.39	32.10.34 A	0.15
23	7.	17.38.2	264.30.24	34.40.34 A	0.30
23	8.	17.49.58	267.29.30	24.21.10 A	0.30
28	9.	17.5.22	256.20.36	18.13.26 A	0.3
29	10.	16.44.48	251.12.6	3.42.18 A	0.4

Détails des Nébuleuses & des amas d'Étoiles.
Les positions sont rapportées ci-contre.

très-bien, par un beau ciel, avec une lunette ordinaire d'un pied. M. Messier l'a rapportée sur la Carte de la Comète de 1763. *Mém. Acad. année 1774, page 40.* Revue les 5 Sept. 1780, 30 Janvier & 22 Mars 1781.

6. Amas de petites étoiles entre l'arc du Sagittaire & la queue du Scorpion. À la vue simple, cet amas semble former une nébulosité sans étoile; mais avec le moindre instrument que l'on emploie pour l'examiner on y voit un amas de petites étoiles.

7. Amas d'étoiles plus considérable que le précédent; cet amas paroît à la vue simple comme une nébulosité, il est peu éloigné du précédent, placée ntre l'arc du Sagittaire & la queue du Scorpion.

8. Amas d'étoiles qui paroît sous la forme de nébuleuse en le regardant avec une lunette ordinaire de trois pieds; mais avec un excellent instrument on n'y remarque qu'une grande quantité de petites étoiles; auprès de cet amas est une étoile assez brillante, environnée d'une lumière très-foible; c'est la neuvième étoile du Sagittaire, de septième grandeur, suivant Flamstéed: cet amas paroît sous une forme alongée qui s'étend du Nord-est au Sud-ouest, entre l'arc du Sagittaire & le pied droit d'*Ophiucus*.

9. Nébuleuse, sans étoile, dans la jambe droite d'*Ophiucus*: elle est ronde & sa lumière foible. Revue le 22 Mars 1781.

10. Nébuleuse, sans étoile, dans la ceinture d'*Ophiucus*, près de la trentième étoile de cette constellation, sixième grandeur suivant Flamstéed. Cette nébuleuse est belle & ronde; on ne pouvoit la voir que difficilement avec une lunette ordinaire de trois pieds. M. Messier l'a rapportée sur la seconde Carte de la route de la Comète de 1769. *Mém. Acad. année 1775, planche IX.* Revue le 6 Mars 1781.

X

DATE des OBSERVATIONS.	Numéros des Nébuleuses	ASCENSION DROITE.		DÉCLINAISON.	Diamètre en degrés & min.
		En Temps. H. M. S.	En Degrés. D. M. S.	D. M. S.	D. M
. Mai 30	11.	18.30:23	279.35.43	6.31. 1A	0. 4
30	12.	16.34.53	248.43.10	2.30.28A	0. 3
Juin. 1	13.	16.33.15	248.18.48	36.54.44B	0. 6
1	14.	17.25.14	261.18.29	3. 5.45A	0. 7

Détails des Nébuleuses & des amas d'Étoiles.
Les positions sont rapportées ci-contre.

N.° des Nébul.	
11.	Amas d'un grand nombre de petites étoiles, près de l'étoile K d'Antinoüs, que l'on ne voit qu'avec de bons instrumens; avec une lunette ordinaire de trois pieds elle ressemble à une Comète: cet amas est mêlé d'une lumière foible; dans cet amas il y a une étoile de 8.e grandeur. M. Kirch l'observa en 1681. Transact. Philos. n.° 347, page 390. Il est rapporté sur le grand Atlas anglois.
12.	Nébuleuse découverte dans le Serpent, entre le bras & le côté gauche d'Ophiucus: cette nébuleuse ne contient aucune étoile, elle est ronde & sa lumière foible; près de cette nébuleuse est une étoile de la neuvième grandeur. M. Messier l'a rapportée sur la seconde Carte de la Comète observée en 1769. Mém. Acad. 1775, pl. IX. Revue le 6 Mars 1781.
13.	Nébuleuse sans étoile, découverte dans la ceinture d'Hercule: elle est ronde & brillante, le centre plus clair que les bords, on l'aperçoit avec une lunette d'un pied; elle est de deux étoiles, l'une & l'autre de la 8.e grandeur, l'une au-dessus & l'autre au-dessous: la nébuleuse a été déterminée en la comparant à s d'Hercule. M. Messier l'a rapportée sur la Carte de la Comète de 1779, insérée dans les Mémoires de l'Académie, de l'année 1784. Vue par Halley en 1714. Revue les 5 & 30 Janv. 1781. Elle est rapportée sur l'Atlas céleste anglois.
14.	Nébuleuse sans étoile, découverte dans la draperie qui passe par le bras droit d'Ophiucus, & placée sur le parallèle de ζ du Serpent; cette nébuleuse n'est pas grande, sa lumière est foible, on peut la voir cependant avec une lunette ordinaire de trois pieds & demi; elle est ronde. près d'elle est une petite étoile de la neuvième grandeur; sa position a été déterminée en la comparant à γ d'Ophiucus, & M. Messier a rapporté sa position sur la Carte de la Comète de 1769. Mémoires de l'Académie, année 1775, planche IX. Revue le 22 Mars 1781.

DATE des OBSERVATIONS.	Numéros des Nébuleuses	ASCENSION DROITE.		DÉCLINAISON.	Diamètre en degrés & min.
		En Temps. H. M. S.	En Degrés. D. M. S.	D. M. S.	D. M.
1764. Juin. 3	15.	21.18.41	319.40.19	10.40. 3B	0. 3
3	16.	18. 5. 0	271.15. 3	13.51.44A	0. 8
3	17.	18. 7. 3	271.45.48	16.14.44A	0. 5
3	18.	18. 6.16	271.34. 3	17.13.14A	0. 5
5	19.	16.48. 7	252. 1.45	25.54.46A	0. 3
5	20.	17.48.16	267. 4. 5	22.59.10A	

Détails des Nébuleuses & des amas d'Étoiles.
Les positions sont rapportées ci-contre.

N.° des Nébul.	
15.	Nébuleuse sans étoile, entre la tête de Pégase & celle du petit Cheval; elle est ronde, le centre en est brillant, sa position déterminée en la comparant à A du petit Cheval. M. Maraldi, dans les Mémoires de l'Académie de 1746, parle de cette nébuleuse: « J'ai aperçu dit-il, entre l'étoile e de Pégase & β du petit Cheval, « une étoile nébuleuse assez claire, qui est composée de « plusieurs étoiles; son ascension droite est de 319d 27' « 6", & sa déclinaison septentrionale de 11d 2' 22". »
16.	Amas de petites étoiles, mêlé d'une foible lumière, près de la queue du Serpent, à peu de distance du parallèle de ζ de cette constellation; avec une foible lunette cet amas paroît sous la forme d'une nébuleuse.
17.	Traînée de lumière sans étoiles, de cinq à six minutes d'étendue, en forme de fuseau, & à peu-près comme celle de la ceinture d'Andromède, mais d'une lumière très-foible; il y a deux étoiles télescopiques auprès & placées parallèlement à l'Équateur. Par un beau ciel on aperçoit très-bien cette nébuleuse avec une lunette ordinaire de trois pieds & demi. Revue le 22 Mars 1781.
18.	Amas de petites étoiles, un peu au-dessous de la nébuleuse ci-dessus, n.° 17, environné d'une légère nebulosité, cet amas moins apparent que le précédent, n° 16: avec une lunette ordinaire de trois pieds & demi, cet amas paroît sous la forme d'une nébuleuse; mais avec une bonne lunette on n'y voit que des étoiles.
19.	Nébuleuse sans étoiles, sur le parallèle d'Antarès, entre le Scorpion & le pied droit d'Ophiucus: cette nébuleuse est ronde, on la voyoit très-bien avec une lunette ordinaire de trois pieds & demi; l'étoile connue la plus voisine de cette nébuleuse est la vingt-huitième d'Ophiucus, 6.e grandeur, suiv. Flamstéed. Revue le 22 Mars 1781.
20.	Amas de petites étoiles, un peu au-dessus de l'Écliptique, entre l'arc du Sagittaire & le pied droit d'Ophiucus. Revu le 22 Mars 1781.

DATE des OBSERVATIONS.	Numéros des Nébuleuses	ASCENSION DROITE. En Temps. H. M. S.	En Degrés. D. M. S.	DÉCLINAISON. D. M. S.	Diamètre en degrés & min. D. M.
1764. Juin. 5	21.	17.50. 7	267.31.35	22.31.25A	
5	22.	18.21.55	275.28.39	24. 6.11A	0. 6
20	23.	17.42.51	265.42.50	18.45.55A	0.15
20	24.	18. 1.44	270.26. 0	18.26. 0A	1.30
20	25.	18.17.40	274.25. 0	19. 5. 0A	0.10

N.os des Nébul.	Détails des Nébuleuses & des amas d'Étoiles. *Les positions sont rapportées ci-contre.*
21.	Amas d'étoiles, près du précédent; l'étoile connue la plus voisine de ces deux amas est la onzieme du Sagittaire, septieme grandeur, suivant Flamstéed. Les étoiles de ces deux amas sont de la huitieme à la neuvieme grandeur, environnées de nébulosité.
22.	Nébuleuse, au-dessous de l'Écliptique, entre la tête & l'arc du Sagittaire, près d'une étoile de la septieme grandeur, la vingt-cinquieme du Sagittaire, suivant Flamstéed, cette nébuleuse est ronde, ne contient aucune étoile, & on la voit très-bien avec une lunette ordinaire de trois pieds & demi; l'étoile λ du Sagittaire a servi à la détermination. Abraham Ihle, Allemand, la découvrit en 1665, en observant Saturne. M. le Gentil l'observa en 1747, & il en a fait graver la figure. *Mémoires de l'Académie, année 1759, page 470.* Revue le 22 Mars 1781: elle est rapportée sur l'*Atlas* anglois.
23.	Amas d'étoiles, entre l'extrémité de l'arc du Sagittaire & le pied droit d'*Ophiucus*, très-près de l'étoile 65.e d'*Ophiucus*, suivant Flamstéed. Les étoiles de l'amas sont très-près les unes des autres. Sa position déterminée par μ du Sagittaire.
24.	Amas d'étoiles au parallèle du précédent & près de l'extrémité de l'arc du Sagittaire, dans la voie lactée: grande nébulosité dans laquelle il y a plusieurs étoiles de différentes grandeurs: la lumière qui est répandue dans cet amas est divisée en plusieurs parties; c'est le milieu de cet amas qui a été déterminé.
25.	Amas de petites étoiles dans le voisinage des deux amas précédens, entre la tête & l'extrémité de l'arc du Sagittaire: l'étoile connue la plus voisine de cet amas est l'étoile 21.e du Sagittaire, 6.e grandeur, suivant Flamstéed. Les étoiles de cet amas se voient difficilement avec une lunette ordinaire de trois pieds; on n'aperçoit aucune nébulosité. Sa position a été connue par l'étoile μ du Sagittaire.

DATE des OBSERVATIONS.	Numéros des Nébuleuses	ASCENSION DROITE. En Temps. H. M. S.	En Degrés. D. M. S.	DÉCLINAISON. D. M. S.	Diamètre en degrés & min. D. M.
1764.Juin.20	26.	18.32.22	278. 5.25	9.38.14A	0. 2.
Juillet 12	27.	19.49.27	297.21.41	22. 4. 0B	0. 4
27	28.	18. 9.58	272.29.30	24.57.11A	0. 2
29	29.	20.15.38	303.54.29	37.11.57B	
Août. 3	30.	21.27. 5	321.46.18	24.19. 4A	0. 2
3	31.	0.29.46	7.26.32	39. 9.32B	0.40

N.os des Nébul.	Détails des Nébuleuses & des amas d'Étoiles. *Les positions sont rapportées ci-après.*
26.	Amas d'étoiles près des étoiles π & o d'*Antinoüs*, entre lesquelles il y en a une qui a plus de lumière: avec une lunette de trois pieds on ne peut pas les distinguer, il faut y employer un bon instrument. Cet amas ne contient aucune nébulosité.
27.	Nébuleuse sans étoile, découverte dans le Renard, entre les deux pattes de devant, & très-près de l'étoile 14.e de cette constellation, 5.e grandeur suivant Flamstéed; on la voit bien avec une lunette ordinaire de 3 pieds & demi: elle paroît sous une forme ovale, & ne contient aucune étoile. M. Messier en a rapporté la position sur la Carte de la Comète de 1779, qui sera gravée pour le volume de l'Acad. de la même année. Revue le 31 Janvier 1781.
28.	Nébuleuse découverte dans la partie supérieure de l'arc du Sagittaire à un degré environ de l'étoile λ & peu éloignée de la belle nébuleuse qui est entre la tête & l'arc. Elle ne contient aucune étoile; elle est ronde, elle ne peut se voir que difficilement avec une lunette ordinaire de 3 pieds ½. Sa position a été déterminée par λ du Sagittaire. Revue le 20 Mars 1781.
29.	Amas de sept ou huit étoiles très-petites, qui sont au-dessous de γ du Cygne, que l'on voit avec une lunette ordinaire de 3 pieds & demi sous la forme d'une nébuleuse. Sa position déterminée par γ du Cygne. Cet amas est rapporté sur la Carte de la Comète de 1779.
30.	Nébuleuse découverte au-dessous de la queue du Capricorne, très-près de l'étoile 41.e de cette constellation, 6.e grandeur, suivant Flamstéed. On la voit difficilement avec une lunette ordinaire de 3 pieds ½. Elle est ronde & ne contient aucune étoile; sa position déterminée par ζ du Capricorne, M. Messier l'a rapportée sur la Carte de la Comète de 1759. *Mém. Acad. 1760, pl. II.*
31.	La belle nébuleuse de la ceinture d'Andromède, en forme de fuseau; M. Messier l'a examinée avec différens instru-

DATE des Observations.	Numéros des Nébuleuses	ASCENSION DROITE. En Temps. H. M. S.	En Degrés. D. M. S.	DÉCLINAISON. D. M. S.	Diamètre en degrés & min. D. M.
1764. Août. 3	32.	0.29.50	7.27.32	38.45.34 B	0. 2
25	33.	1.40.37	20. 9.17	29.32.25 B	0.15
25	34.	2.27.27	36.51.37	41.39.32 B	0.15
30	35.	5.54.41	88.40. 9	24.33.30 B	0.20

Détails des Nébuleuses & des amas d'Étoiles.
Les positions sont rapportées ci-contre.

mens, & il n'y a reconnu aucune étoile : elle ressemble à deux cônes ou pyramides de lumière opposés par leur base, dont l'axe est dans la direction du Nord-ouest au Sud-est ; les deux pointes de lumières ou les deux sommets étoient à peu-près éloignés l'un de l'autre de 40 minutes de degré ; la base commune des deux pyramides de 15 minutes. Cette nébuleuse fut découverte en 1612, par Simon *Marius*, & observée ensuite par différens Astronomes. M. le Gentil en a donné un dessin dans les Mémoires de l'Académie de *1759, page 453*. Elle est rapportée sur l'*Atlas* anglois.

32. Petite nébuleuse sans étoiles, au-dessous & à quelques minutes de celle de la ceinture d'Andromède ; cette petite nébuleuse est ronde, sa lumière plus foible que celle de la ceinture. M. le Gentil la découvrit le 29 Octobre 1749. M. Messier la vit, pour la première fois, en 1757, & il n'y a reconnu aucun changement.

33. Nébuleuse découverte entre la tête du Poisson boréal & le grand Triangle, à peu de distance d'une étoile de 6.e grandeur : la nébuleuse est d'une lumière blanchâtre, d'une densité presqu'égale, cependant un peu plus lumineuse aux deux tiers de son diamètre, & ne contient aucue étoile. On la voit difficilement avec une lunette ordinaire d'un pied. Sa position déterminée en la comparant à α du Triangle Revue le 27 Sept. 1780.

34. Amas de petites étoiles, entre la tête de Méduse & le pied gauche d'Andromède, presque sous le parallèle de γ : avec une lunette ordinaire de 3 pieds on distingue les étoiles. Sa position a été déterminée par β de la tête de Méduse.

35. Amas de très-petites étoiles, près du pied gauche de Castor, à peu de distance des étoiles μ & η de cette constellation. M. Messier a rapporté sa position sur la Carte de la Comète 1770. *Mém. Acad. 1771. pl. VII.* Rapportée sur l'*Atlas* anglois.

DATE des Observations.	Numéros des Nébuleuses	ASCENSION DROITE. En Temps. H. M. S.	En Degrés. D. M. S.	DÉCLINAISON. D. M. S.	Diamètre en degrés & min. D. M.
1764. Sept. 2	36.	5.20.47	80.11.42	34. 8. 6 B	0. 9
2	37.	5.37. 1	84.15.12	32.11.51 B	0. 9
25	38.	5.12.41	78.10.12	36.11.51 B	0.15
Oct. 24	39.	21.23.49	320.57.10	47.25. 0 B	1. 0
24	40.	12.11. 2	182.45.30	59.23.50 B	
1765. Janv. 16	41.	6.35.53	98.58.12	20.33. 0 A	
1769. Mars 4	42.	5.23.59	80.59.40	5.34. 6 A	

Détails des Nébuleuses & des amas d'Étoiles.
Les positions sont rapportées ci-contre.

36. Amas d'étoiles dans le Cocher, près de l'étoile φ : avec une lunette ordinaire de 3 pieds & demi on a de la peine à distinguer les étoiles, l'amas ne contient aucune nébulosité. Sa position déterminée par φ.

37. Amas de petites étoiles, peu éloigné du précédent, sur le parallèle de χ du Cocher ; les étoiles sont plus petites, plus rapprochées & renferment de la nébulosité ; avec une lunette ordinaire de 3 pieds & demi, on a de la peine à voir les étoiles : cet amas est rapporté sur la Carte de la seconde Comète de 1771, *Mém. Acad. 1777.*

38. Amas de petites étoiles dans le Cocher, près de l'étoile σ, peu éloigné des deux amas précédens ; celui-ci est de figure carrée & ne contient aucune nébulosité, si on a soin de l'examiner avec une bonne lunette. Son étendue peut avoir 15 minutes de degré.

39. Amas d'étoiles près de la queue du Cygne ; on les voit avec une lunette ordinaire de 3 pieds & demi.

40. Deux étoiles très-près l'une de l'autre & très-petites, placées à la naissance de la queue de la grande Ourse : on a de la peine à les distinguer avec une lunette ordinaire de 6 pieds. C'est en cherchant la nébuleuse placée au-dessus du dos de la grande Ourse, rapportée dans le livre de la figure des Astres, qui devoit avoir en 1660, 183d 32' 41" d'ascension droite, & 60d 20' 33" de déclinaison boréale, que M. Messier n'a pu voir, qu'il a observé ces deux étoiles.

41. Amas d'étoiles au-dessous de *Sirius*, près de ρ du grand Chien ; cet amas paroît nébuleux avec une lunette ordinaire d'un pied : ce n'est qu'un amas de petites étoiles.

42. Position de la belle nébuleuse de l'épée d'Orion, par l'étoile θ qui y est contenue avec trois autres étoiles plus petites qu'on ne peut voir qu'avec de bons instrumens. M. Messier est entré dans de grands détails sur cette grande nébuleuse ; il en a donné un dessin, fait avec

DATE des Observations.	Numéros des Nébuleuses.	ASCENSION DROITE.		DÉCLINAISON.	Diamètre en degré & min.
		En Temps.	En Degrés.		
		H. M. S.	D. M. S.	D. M. S.	D. M.
1769.Mars. 4	43.	5.24.12	81. 3. 0	5.26.37 A	
4	44.	8. 7.22	126.50.30	20.31.38 B	
4	45.	3.33.48	53.27. 4	23.22.41 B	
1771.Févr.19	46.	7.31.11	112.47.43	14.19. 7 A	
19	47.	7.44.16	116. 3.58	14.50. 8 A	
19	48.	8. 2.24	120.36. 0	1.16.42 A	
19	49	12.17.48	184.26.58	9.16. 9 B	

N.os des Nébul.	Détails des Nébuleuses & des amas d'Étoiles. *Les positions sont rapportées ci-contre.*
	le plus grand soin, qu'on peut voir dans les *Mémoires de l'Académie, année 1771, planche VIII.* Ce fut Huyghens qui la découvrit en 1656 : elle a été observée depuis par un grand nombre d'Astronomes. Rapportée sur l'*Atlas* anglois.
43.	Position de la petite étoile qui est environnée de nébulosités & qui est au-dessous de la nébuleuse de l'épée d'Orion. M. Messier l'a rapporté sur le dessin de la grande.
44.	Amas d'étoiles connues sous le nom de nébuleuses du Cancer, la position rapportée est celle de l'étoile C.
45.	Amas d'étoiles, connues sous le nom des *Pléiades.* La position rapportée est celle de l'étoile *Alcyone.*

Fin du Catalogue imprimé de M. Messier.

Ce qui suit a été observé par M. Messier, depuis l'impression de son Mémoire.

46.	Amas de très-petites étoiles, entre la tête du grand Chien & les deux pattes de derrière de la Licorne, déterminé en comparant cet amas à la 2.e étoile du Navire, 6.e grandeur, suivant Flamstéed ; on ne peut voir ces étoiles qu'avec une bonne lunette ; l'amas contient un peu de nébulosité.
47.	Amas d'étoiles peu éloigné du précéd. les étoiles plus grandes ; le milieu de l'amas comparé à la même étoile, la seconde du Navire. L'amas ne contient aucune nébulosité.
48.	Amas de très-petites étoiles, sans nébulosité ; cet amas est à peu de distance des trois étoiles qui sont à la naissance de la queue de la Licorne.
49.	Nébuleuse découverte près de l'étoile ρ de la Vierge. Ce n'est pas sans peine qu'on peut la voir avec une lunette ordinaire de 3 pieds & demi. La Comète de 1779 fut comparée par M. Messier à cette nébuleuse les 22 & 23 Avril : la Comète & la Nébuleuse avoient même

DATE des Observations.	Numéros des Nébuleuses.	ASCENSION DROITE.		DÉCLINAISON.	Diamètre en degré & min.
		En Temps.	En Degrés.		
		H. M. S.	D. M. S.	D. M. S.	D. M.
1772. Avril. 5	50.	6.51.50	102.57.28	7.57.42 A	
1774.Janv.11	51.	13.20.23	200. 5.48	48.24.24 B	
1774.Sept. 7	52.	23.14.38	348.39.27	60.22.12 B	
1777.Févr.26	53.	13. 2. 2	195.30.26	19.22.44 B	

N.os des Nébul.	Détails des Nébuleuses & des amas d'Étoiles. *Les positions sont rapportées ci-contre.*
	lumière. M. Messier a rapporté cette nébuleuse sur la Carte de la route de cette Comète, qui paroîtra dans le volume de l'Académie de la même année 1779. Revue le 10 Avril 1781.
50.	Amas de petites étoiles plus ou moins brillantes, au-dessous de la cuisse droite de la Licorne, au-dessus de l'étoile θ de l'oreille du grand Chien, & près d'une étoile de 7.e grandeur. C'est en observant la Comète de 1772 que M. Messier observa cet amas. Il l'a rapporté sur sa Carte de cette Comète, qu'il en a tracée. *Mém. Acad. 1772.*
51.	Nébuleuse très-foible, sans étoiles, près de l'oreille des Lévriers, la plus septentrionale, au-dessous de l'étoile η 2.e grandeur de la queue de la grande Ourse : M. Messier découvrit cette nébuleuse le 13 Octobre 1773, en observant la Comète qui parut cette année. On ne peut la voir que difficilement avec une lunette ordinaire de 3 pieds ½ : près d'elle est une étoile de la 8.e grandeur. M. Messier a rapporté sa position sur la Carte de la Comète observée en 1773 & 1774. *Mémoires de l'Académie 1774, planche III.* Elle est double, ayant chacune un centre brillant, éloigné l'un de l'autre de 4′ 35″. Les deux atmosphères se touchent. L'une est plus foible que l'autre. Revue plusieurs fois.
52.	Amas de très-petites étoiles, mêlé de nébulosité, qu'on ne peut voir qu'avec une lunette achromatique. C'est en observant la Comète qui parut cette année que M. Messier vit cet amas qui étoit près de la Comète le 7 de Septembre 1774 ; il est au-dessous de l'étoile δ de Cassiopée : cette étoile δ servit à déterminer l'amas d'étoiles & la Comète.
53.	Nébuleuse sans étoiles, découverte au-dessous & près de la chevelure de Bérénice, à peu de distance de l'étoile 42.e de cette constellation, suivant Flamstéed. Cette nébuleuse est ronde & apparente. La Comète de 1779 fut comparée directement à cette nébuleuse, & M. Messier l'a

DATE des OBSERVATIONS.	Numéros des Nébuleuses	ASCENSION DROITE. En Temps. H. M. S.	ASCENSION DROITE. En Degrés. D. M. S.	DÉCLINAISON. D. M. S.	Diamètre en degrés & min. D. M.
1778. Juill. 24	54	18. 40. 52	280. 12. 55	30. 44. 1 A	
24	55	19. 26. 2	291. 30. 25	31. 26. 27 A	
1779. Janv. 23	56	19. 8. 0	287. 0. 1	29. 48. 14 B	
31	57	18. 45. 21	281. 20. 8	32. 46. 3 B	

N.ºˢ des Nébul.	Détails des Nébuleuses & des amas d'Étoiles. *Les positions sont rappportées ci-contre.*
	rapportée sur la Carte de cette Comète, qui sera inférée dans le volume de l'Académie de 1779. Revue le 13 Avril 1781 : elle ressemble à la nébuleuse qui est au-dessous du Lièvre.
54.	Nébuleuse très-foible, découverte dans le Sagittaire ; le centre en est brillant & ne contient aucune étoile, vue avec une lunette achromatique de 3 pieds ½. Sa position a été déterminée par ζ du Sagittaire 3.ᵉ grandeur.
55.	Nébuleuse qui est une tache blanchâtre, de 6 minutes environ d'étendue, sa lumière est égale & n'a paru contenir aucune étoile. Sa position a été déterminée par ζ du Sagittaire, au moyen d'une étoile intermédiaire de 7.ᵉ grandeur. Cette nébuleuse avoit été découverte par M. l'abbé de la Caille. *Mém. Acad. 1755, p. 194.* M. Messier l'avoit cherchée inutilement le 29 Juillet 1764, comme il le rapporte dans son Mémoire.
56.	Nébuleuse sans étoile, ayant peu de lumière ; M. Messier la découvrit le jour même de la découverte de la Comète de 1779 le 19 Janvier. Le 23 il en détermina sa position en la comparant à l'étoile n.ᵉ 2 du Cygne, suivant Flamstéed : elle est près de la voie lactée ; auprès d'elle est une étoile de la dixième grandeur. M. Messier l'a rapportée sur la Carte de la Comète de 1779.
57.	Amas de lumière placé entre γ & β de la Lyre, découvert en observant la Comète de 1779, qui en a passé très-près : il sembloit que cet amas de lumière, qui est arrondi, étoit composé de très-petites étoiles : avec les meilleures lunettes il n'est pas possible de les appercevoir, il reste seulement un soupçon qu'il y en a. M. Messier a rapporté cet amas de lumière sur la Carte de la Comète de 1779. M. Darquier, à Toulouse, découvrit cette nébuleuse, en observant la même Comète, & il rapporte : « Nébuleuse entre γ & β de la Lyre ; elle est fort terne, « mais parfaitement terminée ; elle est grosse comme « Jupiter & ressemble à une Planète qui s'éteindroit ».

DATE des OBSERVATIONS.	Numéros des Nébuleuses	ASCENSION DROITE. En Temps. H. M. S.	ASCENSION DROITE. En Degrés. D. M. S.	DÉCLINAISON. D. M. S.	Diamètre en degrés & min. D. M.
1779. Avril 15	58	12. 26. 30	186. 37. 23	13. 2. 42 B	
15	59	12. 30. 47	187. 41. 38	12. 52. 36 B	
15	60	12. 32. 28	188. 6. 53	12. 46. 2 B	
Mai. 11	61	12. 10. 44	182. 41. 5	5. 42. 5 B	
1779. Juin. 4	62	16. 47. 14	251. 48. 24	29. 45. 30 A	

N.ºˢ des Nébul.	Détails des Nébuleuses & des amas d'Étoiles. *Les positions sont rapportées ci-contre.*
58.	Nébuleuse très-foible découverte dans la Vierge, presque sur la parallèle de ε 3.ᵉ grandeur. La moindre lumière pour éclairer les fils du micromètre la faisoit disparoître. M. Messier l'a rapportée sur la Carte de la Comète de 1779, qui se trouvera dans le volume de l'Académie de la même année.
59.	Nébuleuse dans la Vierge & dans le voisinage de la précédente, sur la parallèle de ε, qui a servi à sa détermination : elle est de la même lumière que celle ci-dessus, aussi foible. M. Messier l'a rapportée sur la Carte de la Comète de 1779.
60.	Nébuleuse dans la Vierge, un peu plus apparente que les deux précédentes, de même sur la parallèle de ε, qui a servi à sa détermination. M. Messier l'a rapportée sur la Carte de la Comète de 1779. Il découvrit ces trois nébuleuses en observant cette Comète qui passa très-près d'elles. La dernière en passa si près les 13 & 14 Avril, qu'étant l'une & l'autre dans le champ de la lunette, il ne put la voir ; ce ne fut que le 15, en cherchant la Comète, qu'il apperçut une nébuleuse. Ces trois nébuleuses ne paroissoient contenir aucune étoile.
61.	Nébuleuse très-foible & difficile à appercevoir. M. Messier prit cette nébuleuse pour la Comète de 1779, les 5, 6 & 11 Mai ; le 11 il reconnut que ce n'étoit pas la Comète ; mais une nébuleuse qui se trouvoit sur sa route & au même point du Ciel.
62.	Nébuleuse très-belle, découverte dans le Scorpion, elle ressemble à une petite Comète, le centre en est brillant & environné d'une lumière foible. Sa position déterminée, en la comparant à l'étoile τ du Scorpion. M. Messier avoit déja vu cette nébuleuse le 7 Juin 1771, sans en avoir déterminé le lieu qu'à peu-près. Revue le 22 Mars 1781.

DATE des Observations.	Numéros des Nébuleuses.	ASCENSION DROITE.		DÉCLINAISON.	Diamètre en degrés & min.
		En Temps.	En Degrés.		
		H. M. S.	D. M. S.	D. M. S.	D. M.
1779.Juin. 14	63.	13. 4. 22	196. 5. 30	43. 12. 37 B	
1780. Mars 1	64.	12. 45. 51	191. 27. 38	22. 52. 31 B	
1	65.	11. 7. 24	166. 50. 54	14. 16. 8 B	
1	66.	11. 8. 47	167. 11. 39	14. 12. 21 B	
Avril. 6	67.	8. 36. 28	129. 6. 57	12. 36. 38 B	
9	68.	12. 27. 38	186. 54. 33	25. 30. 20 A	

N.os des Nébul.	Détails des Nébuleuses & des amas d'Étoiles. *Les positions sont rapportées ci-contre.*
63.	Nébuleuse découverte par M. Méchain dans les Chiens de chasse. M. Messier l'a cherchée ; elle est foible , elle a à peu-près la même lumière que la nébuleuse rapportée sous le n.º 59 : elle ne contient aucune étoile, & la moindre lumière pour éclairer les fils du micromètre la faisoit disparoître : il y a auprès d'elle une étoile de la 8.º grandeur, qui précède la nébuleuse au fil horaire. M. Messier en a rapporté la position sur la Carte de la route de la Comète de 1779.
64.	Nébuleuse découverte dans la chevelure de Bérénice, qui est moins apparente de moitié que celle qui est au-dessous de la chevelure. M. Messier en a rapporté la position sur la Carte de la Comète de 1779. Révue le 17 Mars 1781.
65.	Nébuleuse découverte dans le Lion ; elle est très-foible & ne contient aucune étoile.
66.	Nébuleuse découverte dans le Lion ; sa lumière très-foible & très-près de la précédente : elles paroissent l'une & l'autre dans le même champ de la lunette. La Comète observée en 1773 & 1774 avoit passé entre ces deux nébuleuses du 1 au 2 Novembre 1773. M. Messier ne les vit pas alors , sans doute , à cause de la lumière de la Comète.
67.	Amas de petites étoiles avec de la nébulosité , au-dessous de la Serre australe de l'Écrevisse. La position déterminée par l'étoile α.
68.	Nébuleuse sans étoiles au-dessous du Corbeau & de l'Hydre ; elle est très-foible , très-difficile à apercevoir avec les lunettes ; près d'elle est une étoile de la sixième grandeur.

DATE des Observations.	Numéros des Nébuleuses.	ARCENSION DROITE.		DÉCLINAISON.	Diamètre en degrés & min.
		En Temps.	En Degrés.		
		H. M. S.	D. M. S.	D. M. S.	D. M.
1780. Août 31	69.	18. 16. 47	274. 11. 46	32. 31. 45 A	0. 2
31	70.	18. 28. 53	277. 13. 16	32. 31. 7 A	0. 2
Octob. 4	71.	19. 43. 57	295. 59. 9	18. 13. 0 B	0. 3 ½
M. Méchain.	...	....	296. 0. 4	18. 14. 21 B	

N.os des Nébul.	Détails des Nébuleuses & des amas d'Étoiles. *Les positions sont rapportées ci-contre.*
	ADDITION au Catalogue des Nébuleuses & des amas d'Étoiles de M. Messier, inséré dans la Connoiss. des Temps de 1783, page 225, & année 1784, page 255 & suivantes.
69.	Nébuleuse sans étoiles, dans le Sagittaire. au-dessous de son bras gauche & près de l'arc ; près d'elle est une étoile de la 9.º grandeur : sa lumière très-foible, on ne peut la voir que par un beau temps, & la moindre lumière employée pour éclairer les fils du micromètre la faisoit disparoître : sa position a été déterminée par ε du Sagittaire : cette nébuleuse a été observée par M. de la Caille, & rapportée dans son Catalogue ; elle ressemble au noyau d'une petite Comète.
70.	Nébuleuse sans étoile, près de la précédente, & sur le même parallèle : près d'elle est une étoile de la neuvième grandeur & quatre petites étoiles télescopiques, presque sur une même ligne droite, très-près les unes des autres, & sont placées au-dessus de la nébuleuse, vue dans une lunette qui renverse ; la nébuleuse déterminée par la même étoile ε du Sagittaire.
71.	Nébuleuse découverte par M. Méchain le 28 Juin 1780, entre les étoiles γ & δ de la Flèche. Le 4 Octobre suivant, M. Messier l'a cherchée : sa lumière est très-foible & ne contient aucune étoile ; la moindre lumière la faisoit disparoître. Elle est placée à environ 4 degrés au-dessous de celle que M. Messier découvrit dans le Renard. Voyez n.º 27. Il la rapporte sur la Carte de la Comète de 1779.

DATE des Observations.	Numéros des Nébuleuses.	ASCENSION DROITE.		DÉCLINAISON.	Diamètre en degrés & min.
		En Temps.	En Degrés.		
		H. M. S.	D. M. S.	D. M. S.	D. M.
1780. Oct. 4. M. Méchain.	72. ...	20.41.23	310.20.49 / 310.21.10	13.20.51A / 13.21.24	0. 2
Octob. 4 & 5	73. ...	20.46.52	311.43. 4	13.28.40A	
18 M. Méchain.	74. ...	1.24.57	21.14. 9 / 21.17. 0	14.39.35 B / 14.36. 0	
18 M. Méchain.	75. ...	19.53.10	298.17.24 / 298.17.30	22.32.23A / 22.32. 0	
21 M. Méchain.	76. ...	1.28.43	22.10.47 / 22.10.26	50.28.48B / 50.28.12	0. 2

N.os des Nébul.	Détails des Nébuleuses & des amas d'Étoiles. Les positions sont rapportées ci-contre.
72.	Nébuleuse vue par M. Méchain la nuit du 29 au 30 Août 1780, au dessus du cou du Capricorne. M. Messier l'a cherchée les 4 & 5 Octobre suivant : sa lumière foible comme la précédente ; près d'elle est une petite étoile télescopique : sa position fut déterminée par l'étoile ν du Verseau, cinquième grandeur.
73.	Amas de trois ou quatre petites étoiles, qui ressemble à une nébuleuse au premier coup-d'œil, contient un peu de nébulosité : cet amas est placé sur le parallèle de la nébuleuse précédente : sa position a été déterminée par la même étoile ν du Verseau.
74.	Nébuleuse sans étoile, près de l'étoile η du Lien des Poissons, vue par M. Méchain à la fin de Septembre 1780, & qu'il rapporte. « Cette nébuleuse ne contient « pas d'étoiles ; elle est assez large, très-obscure, extrê- « mement difficile à observer, on pourra la déterminer « plus exactement dans les belles gelées ». M. Messier l'a recherchée & l'a trouvée, comme l'a décrit M. Méchain : elle a été comparée directement à l'étoile η des Poissons.
75.	Nébuleuse sans étoile, entre le Sagittaire & la tête du Capricorne : vue par M. Méchain le 27 & le 28 Août 1780. M. Messier l'a cherchée le 5 Octobre suivant & le 18 l'a comparée à l'étoile n.º 4, sixième grandeur du Capricorne, suivant Flamstéed : il a semblé à M. Messier qu'elle n'étoit composée que de très-petites étoiles, contenant de la nébulosité : M. Méchain l'a rapportée comme nébuleuse sans étoiles. M. Messier la vit le 5 Octobre ; mais la Lune étoit sur l'horizon, & ce ne fut que le 18 du même mois qu'il put juger de ses apparences & en déterminer son lieu.
76.	Nébuleuse au pied droit d'Andromède, vue par M. Méchain le 5 Sept. 1780, & qu'il rapporte : « Cette Nébuleuse ne contient pas d'étoiles ; elle est petite & foible ». Le

DATE des Observations.	Numéros des Nébuleuses.	ASCENSION DROITE.		DÉCLINAISON.	Diamètre en degrés & min.
		En Temps.	En Degrés.		
		H. M. S.	D. M. S.	H. M. S.	D. M.
1780. Déc. 17. M. Méchain.	77. ...	2.31.30	37.52.33 / 37.52.58	0.57.43A / 0.57.44	
17 M. Méchain.	78. ...	5.35.34	83.53.35 / 83.53. 2	0. 1.23A / 0. 0.31	0. 3
17 M. Méchain.	79. ...	5.15.16	78.49. 2 / 78.47.10	24.42.57A / 24.44.46	
1781. Janv. 4. M. Méchain.	80. ...	16. 4. 0	240.59.48 / 241. 0.26	22.25.13A / 22.27.58	0. 2

N.os des Nébul.	Détails des Nébuleuses & des amas d'Étoiles. Les positions sont rapportées ci-contre.
	21 Octobre suivant M. Messier la chercha avec sa lunette achromatique, & il lui a semblé qu'elle n'étoit composée que de très-petites étoiles, qui contenoit de la nébulosité, & que la moindre lumière employée pour éclairer les fils du micromètre les faisoient disparoître : la position déterminée par l'étoile φ d'Andromède, quatrième grandeur.
77.	Amas de petites étoiles, qui contient de la nébulosité, dans la Baleine, & sur le parallèle de l'étoile δ rapportée, de troisième grandeur, & que M. Messier n'a estimé que de la cinquième. M. Méchain vit cet amas le 9 Octobre 1780 sous la forme de nébuleuse.
78.	Amas d'étoiles, avec beaucoup de nébulosité dans Orion & sur le parallèle de l'étoile du Baudrier, qui a servi à en déterminer son lieu ; l'amas suivit l'étoile au fil horaire de 3ᵈ 41′ & l'amas supérieur à l'étoile de 27′ 7″. M. Méchain avoit vu cet amas au commencement de 1780, & le rapporte ainsi : « Sur le côté gauche d'Orion, 2 à 3 minutes de diamètre, on « y voit deux noyaux assez brillans, entourés d'une « nébulosité. »
79.	Nébuleuse sans étoile, placée au-dessous du Lièvre, & sur le parallèle d'une étoile de la sixième grandeur : vue par M. Méchain le 26 Octobre 1780. M. Messier la chercha le 17 Décembre suivant : cette nébuleuse est belle ; le centre brillant, la nébulosité peu diffuse : sa position déterminée par l'étoile ε du Lièvre, quatrième grandeur.
80.	Nébuleuse sans étoile, dans le Scorpion, entre les étoiles g & δ, comparée à g pour en déterminer son lieu : cette nébuleuse est ronde, le centre brillant & ressemble à un noyau d'une petite Comète, environné de nébulosité. M. Méchain la vit le 27 Janvier 1781.

DATE des OBSERVATIONS.	Numéros des Nébuleuses	ASCENSION DROITE.		DÉCLINAISON.	Diamètres en degrés & min.
		En Temps.	En Degrés.		
		H. M. S.	D. M. S.	D. M. S.	D. M.
1781. Févr. 9	81.	9.37.51	144.27.44	70. 7.24 B	
M. Méchain.	...	....	144.27. 0	70. 4. 0	
9	82.	9.37.57	144.29.22	70.44.27 B	
M. Méchain.	...		144.28.13	70.43. 5	
17	83.	13.24.33	201. 8.13	28.42.27 A	
Mars 18	84.	12.14. 1	183.30.21	14. 7. 1 B	
18	85.	12.14.21	183.35.21	19.24.26 B	
M. Méchain.	...	....	183.35.45	19.23. 0	

N.os des Nébul.	Détails des Nébuleuses & des amas d'Étoiles. *Les positions sont rapportées ci-contre.*
81.	Nébuleuse près de l'oreille de la grande Ourse, sur le parallèle de l'étoile *d*, de la quatrième à la cinquième grandeur : sa position déterminée par cette étoile. Cette nébuleuse est un peu ovale, le centre clair, & on la voit très-bien avec une lunette ordinaire de trois pieds & demi. Elle fut découverte à Berlin, par M. Bode, le 31 Décembre 1774, & par M. Méchain, au mois d'Août 1779.
82.	Nébuleuse sans étoile, près de la précédente ; l'une & l'autre paroissoient en même-temps dans le champ de la lunette, celle-ci moins apparente que la précédente ; sa lumière foible & alongée : à son extrémité est une étoile telescopique. Vue à Berlin, par M. Bode, le 31 Décembre 1774, & par M. Méchain au mois d'Août 1779.
83.	Nébuleuse sans étoile, près de la tête du Centaure : elle paroissoit sous une lumière foible & égale, mais si difficile à voir avec la lunette, que la moindre lumière pour éclairer les fils du micromètre la faisoit disparoître. Ce ne sera qu'avec beaucoup d'attention qu'on pourra la voir : elle forme un triangle avec deux étoiles estimées de sixième & de septième grandeur : déterminée par les étoilés *i, k, h*, de la tête du Centaure : M. de la Caille avoit déjà déterminé cette nébuleuse. Voyez à la fin de ce Catalogue.
84.	Nébuleuse sans étoile, dans la Vierge ; le centre en est un peu brillant, environné d'une légère nébulosité : sa lumière & ses apparences ressemble à celles de ce Catalogue, n.os 59 & 60.
85.	Nébuleuse sans étoile, au-dessus & près de l'épi de la Vierge, entre les deux étoiles de la chevelure de Bérénice, n.os 11 & 14 du Catalogue de Flamstéed : cette nébuleuse est très-foible. M. Méchain en avoit déterminé sa position le 4 Mars 1781.

DATE des OBSERVATIONS.	Numéros des Nébuleuses	ASCENSION DROITE.		DÉCLINAISON.	Diamètre en degrés & min.
		En Temps.	En Degrés.		
		H. M. S.	D. M. S.	D. M. S.	D. M.
1781. Mars 18	86.	12.15. 5	183.46.21	14. 9.52 B	
18	87.	12.19.48	184.57. 6	13.38. 1 B	
18	88.	12.21. 3	185.15.49	15.37.51 B	
18	89.	12.24.38	186. 9.36	13.46.49 B	
18	90.	12.25.48	186.27. 0	14.22.50 B	
18	91.	12.26.28	186.37. 0	14.57. 6 B	
18	92.	17.10.32	257.38. 3	43.21.59 B	0. 5

N.os des Nébul.	Détails des Nébuleuses & des amas d'Étoiles. *Les positions sont rapportées ci-contre.*
86.	Nébuleuse sans étoile, dans la Vierge, sur le parallèle & très-près de la nébuleuse ci-dessus, n.° 84 ; ses apparences les mêmes, & l'une & l'autre paroissoient dans le même champ de la lunette.
87.	Nébuleuse sans étoile, dans la Vierge, au-dessous & assez près d'une étoile de huitième grandeur, l'étoile ayant même ascension droite que la nébuleuse, & sa déclinaison étoit de 13 d 42′ 21″ boréale. Cette nébuleuse paroissoit de la même lumière que les deux nébuleuses n.os 84 & 86.
88.	Nébuleuse sans étoile, dans la Vierge, entre deux petites étoiles & une étoile de la sixième grandeur, qui paroissoient en même-temps que la nébuleuse dans le champ de la lunette. Sa lumière est une des plus foibles & ressemble à celle rapportée dans la Vierge, n.° 58.
89.	Nébuleuse sans étoile, dans la Vierge, à peu de distance & sur le parallèle de la Nébuleuse ci-dessus rapportée, n.° 87. Sa lumière étoit extrêmement foible & rare, ce n'est pas sans peine qu'on peut l'appercevoir.
90.	Nébuleuse sans étoile, dans la Vierge : sa lumière aussi foible que la précédente, n.° 89.
91.	Nébuleuse sans étoile, dans la Vierge, au-dessus de la précédente n.° 90 : sa lumière encore plus foible que celle ci-dessus. *Nota.* La constellation de la Vierge, & sur-tout l'aile boréale est une des constellations qui renferme le plus de Nébuleuse : ce Catalogue en contient treize de déterminées : savoir, les n.os 49, 58, 59, 60, 61, 84, 85, 86, 87, 88, 89, 90 & 91. Toutes ces nébuleuses paroissent sans étoiles : on ne pourra les voir que par un très-beau ciel, & vers leurs passages au Méridien. La plupart de ces nébuleuses m'avoient été indiquées par M. Méchain.
92.	Nébuleuse, belle, apparente, & d'une grande lumière, entre le genou & la jambe gauche d'Hercule, se voit

DATE des Observations.	Numéros des Nébuleuses	Ascension droite. En Temps. H. M. S.	Ascension droite. En Degrés. D. M. S.	Déclinaison. D. M. S.	Diamètre en degrés & min. D. M.
1781. Mars 20	93.	7. 35. 14	113. 48. 35	23. 19. 45 A	0. 8
24	94.	12. 40. 43	190. 10. 46	42. 18. 43 B	0. 2½
M. Méchain.	...	....	190. 9. 38	42. 18. 50	
24	95.	10. 32. 12	158. 3. 5	12. 50. 21 B	
M. Méchain.	...	....	158. 6. 23	12. 49. 50	
24	96.	10. 35. 5	158. 46. 20	12. 58. 9 B	
M. Méchain.	...	....	158. 48. 0	12. 57. 33	
24	97.	11. 1. 15	165. 18. 40	56. 13. 30 A	0. 2

Détails des Nébuleuses & des amas d'Étoiles.

Les positions sont rapportées ci-contre.

N.ᵒˢ des Nébul.	
	très-bien avec une lunette d'un pied. Elle ne contient aucune étoile; le centre en est clair & brillant, environné de nébulosité & ressemble au noyau d'une grosse Comète : sa lumière, sa grandeur, approchent beaucoup de la nébuleuse qui est dans la ceinture d'Hercule. Voyez *n.° 13* de ce Catalogue : sa position a été déterminée, en la comparant directement à l'étoile σ d'Hercule, quatrième grandeur : la nébuleuse & l'étoile sur le même parallèle.
93.	Amas de petites étoiles, sans nébulosité, entre le grand Chien & la proue du Navire.
94.	Nébuleuse sans étoile, au-dessus du cœur de Charles, sur le parallèle de l'étoile *n.° 8*, sixième grandeur des Lévriers, suivant Flamstéed : le centre en est brillant & la nébulosité peu diffuse. Elle ressemble à la nébuleuse qui est au-dessous du Lièvre, *n.° 79*; mais celle-ci est plus belle & plus brillante : M. Méchain en fit la découverte le 22 Mars 1781.
95.	Nébuleuse sans étoile, dans le Lion, au-dessus de l'étoile *l*: sa lumière est très-foible.
96.	Nébuleuse sans étoile, dans le Lion, près de la précédente; celle-ci moins apparente, toutes deux sur le parallèle de *Régulus*: elles ressemblent aux deux Nébuleuses de la Vierge, *n.ᵒˢ 84 & 86*. M. Méchain les vit toutes deux le 20 Mars 1781.
97.	Nébuleuse dans la grande Ourse, près de β: elle est difficile à voir, rapporte M. Méchain, sur-tout quand on éclaire les fils du micromètre : sa lumière est foible, sans étoile. M. Méchain la vit pour la première fois le 16 Février 1781, & la position est rapportée d'après lui. Près de cette nébuleuse il en vit une autre, qui n'a pas encore été déterminée, ainsi qu'une troisième qui est auprès de γ de la grande Ourse.

DATE des Observations.	Numéros des Nébuleuses	Ascension droite. En Temps. H. M. S.	Ascension droite. En Degrés. D. M. S.	Déclinaison. D. M. S.	Diamètre en degrés & min. D. M.
1781. Avril 13	98.	12. 3. 23	180. 50. 49	16. 8. 15 B	
13	99.	12. 7. 41	181. 55. 19	15. 37. 12 B	
13	100.	12. 11. 57	182. 59. 19	16. 59. 21 B	
1781. Mars 27	101.	13. 43. 28	208. 52. 4	55. 24. 25 B	0. 7

Voyez année 1784, page 266.

Détails des Nébuleuses & des amas d'Étoiles.

Les positions sont rapportées ci-contre.

N.ᵒˢ des Nébul.	
98.	Nébuleuse sans étoile, d'une lumière extrêmement foible, au-dessus de l'aile boréale de la Vierge, sur le parallèle & près de l'étoile *n.° 6*, cinquième grandeur, de la chevelure de Bérénice, suivant Flamstéed. M. Méchain la vit le 15 Mars 1781.
99.	Nébuleuse sans étoile, d'une lumière très-rare, cependant un peu plus claire que la précédente, placée sur l'aile boréale de la Vierge, & près de la même étoile, *n.° 6*, de la chevelure de Bérénice. La nébuleuse est entre deux étoiles de septième & de huitième grandeur. M. Méchain la vit le 15 Mars 1781.
100.	Nébuleuse sans étoile, de la même lumière que la précédente, placée dans l'épi de la Vierge. Vue par M. Méchain le 15 Mars 1781. Ces trois nébuleuses, *n.ᵒˢ 98, 99 & 100*, sont très-difficiles à reconnoître, à cause de la foiblesse de leurs lumières : on ne pourra les voir que par un beau temps, & vers leurs passages au Méridien.
	Par M. Méchain, que M. Messier n'a pas encore vue.
101.	Nébuleuse sans étoile, très-obscure & fort large, de *6* à *7* minutes de diamètre, entre la main gauche du Bouvier & la queue de la grande Ourse. On a peine à la distinguer en éclairant les fils.
102.	Nébuleuse entre les étoiles ο du Bouvier & ι du Dragon : elle est très-foible; près d'elle est une étoile de la sixième grandeur.
103.	Amas d'étoiles entre ε & δ de la jambe de Cassiopée.

Voyez année 1784, page 267.

How This Album Was Compiled

O N PAGE 33 of this book, our Messier album begins — an amateur observer's guide to most of the finest nebulae, clusters, and galaxies visible from mid-northern latitudes. For each object there is a photograph, often a drawing, a finder chart, and a description of the visual appearance, all from observations made by the authors. Published sources were used only for a paragraph or two of basic data and for the descriptions from the *New General Catalogue* by J. L. E. Dreyer (1888).

THE VISUAL OBSERVATIONS

Between 1958 and 1962, in the course of a more extensive sky survey, John H. Mallas examined all the Messier objects. This work was done at Covina, California, with a 4-inch f/15 Unitron refractor that had an equatorial mount and clock drive. As far as light-gathering power and resolution are concerned, this telescope was no doubt superior to the various small 18th-century instruments employed by Messier.

At Covina, the sky was dark in the east and reasonably so to the north and south. But to the west, even after midnight, there was a bothersome glow on the horizon caused by Los Angeles city lights. When a temperature inversion prevailed, smog existed until after midnight. By necessity, most clusters, nebulae, and galaxies were observed in the second half of the night. Observing was done only when conditions were rated good or excellent for the 4-inch.

John Mallas and his 4-inch Unitron refractor. A computer expert, Mr. Mallas passed away in October, 1975.

The Messier list contains such a variety of objects that no one aperture or magnification can give good views of them all. Hence, published descriptions may mean little unless the size of telescope used is specified. And there is no substitute for personal experience. For example, the Pleiades cluster (M45) is best seen with a 2-inch at about 15x but is disappointing in a 4-inch. However, the nebulosity associated with the Pleiades requires a 4-inch or larger. Similarly, while the globular cluster M13 in Hercules needs a 6-inch telescope to be well resolved, a 4-inch will suffice for M4 and M22 if the observer's location is favorable for viewing these southerly globulars. Some Messier objects show additional detail with each increase in aperture, but others do not. The sometimes subtle differences among these sky objects are a challenge. As for the beginner, he should gain experience with his first instrument, even though it is small, before deciding on other equipment.

The visual descriptions and drawings presented in this album are not conservative. They represent what an experienced amateur might see with an instrument of modest size under good conditions. But since there is a natural tendency for a visual observer to look for details that he remembers from photographs, no drawing of a nebula or galaxy can be quite free from bias. In addition, the eye strives to recognize geometrical patterns in star groupings, and it tends to emphasize some features over others in, say, a nebula.

Most of the drawings are of globulars and galaxies. The sketches were made on vellum-type drafting paper with a soft pencil, using finger smudging and erasing until the desired effects were achieved. Contrasts have been exaggerated in order to include the fainter details. Positions of features in the freehand sketches are less accurate than the portrayals of their visual characteristics, though the drawings were made as accurately as possible.

In many of the drawings, foreground stars have been omitted, as in the view of M81 in Ursa Major, for the observer has concentrated on the subjects themselves. On the other hand, in sketching M42-43 (the Orion nebula), all the BD stars were retained, since they affect the visual impression of the nebula to a great extent. Every bright globular cluster seems bordered by subjective star strings, spider legs, or the like, and this effect was especially retained in the M5 drawing. For the other globulars, medium to high powers were used, showing features seldom apparent at low magnification. In making any drawing with the 4-inch, the general rule was to eliminate foreground stars if possible and to use the power that showed interesting detail.

While observer Mallas had previous knowledge of how *types* of objects appeared photographically, in about two-thirds of the cases he had not seen a photograph of the individual object beforehand. It was only after the survey was complete that he compared all the drawings with photographs. It became apparent that the visual impressions of *galaxies* can be correlated with their photographic properties, and in fact give clues to their Hubble types. The most useful criterion is the size of the bright central region. Visually, an Sa spiral can often be recognized by its large nuclear region, whereas an Sc spiral has a small, almost starlike core. In poor seeing this difference is not apparent. The experienced eye under favorable observing conditions can judge the texture of an object as fine-grained, granular, fluffy, or mottled. Long practice counts greatly in this type of observing.

Only a few drawings of *galactic clusters* are given, as the photographic and visual appearances are very similar. M11 in Scutum is a notable exception, and the sketch of M41 is presented to illustrate how the eye introduces geometrical patterns that are absent in the photograph. The Pleiades (M45), of course, is in a class by itself. The advantage of high power is demonstrated by a comparison of sketches of *planetary nebulae* with photographs. The minute details are not seen at low powers.

THE PHOTOGRAPHIC OBSERVATIONS

Evered Kreimer carried out his photographic work for this album from Prescott, Arizona. There, 5,400 feet above sea level, observing conditions were generally good, but deteriorated as the population grew. Most of the town was south of the observatory, and it was usually necessary to wait until after midnight before the dust and smoke settled, giving a darker sky. Even so, faint objects in the far southern sky can no longer be photographed.

All the negatives were taken with a 12½-inch f/7 Cave reflector, mounted on the top of the Kreimer house. This telescope had big setting circles and push-button electric slow motions in both coordinates (with a right-ascension differential gear). A rotating tube permitted comfortable working in any part of the sky, and a shield at its end gave added protection against street lights.

The camera used is more properly termed a guiding head. Placed at the Newtonian focus, it consisted of a base ring, a focusing ring, and a rotating plate within the latter. The plate held the film and the guiding microscope. In simplicity of operation this arrangement left little to be desired. Most of the photographs were made with the low-temperature camera. Others were made with an earlier cooled camera, and some by conventional techniques.

Comfort during a photographic session was necessary for good guiding on picture after picture. An adjustable observing chair helped bring the observer's eye to the level of the guiding eyepiece, while a box attached to the chair kept filmholders and tools within handy reach. It was not difficult to average eight pictures during a four-hour observing session.

30

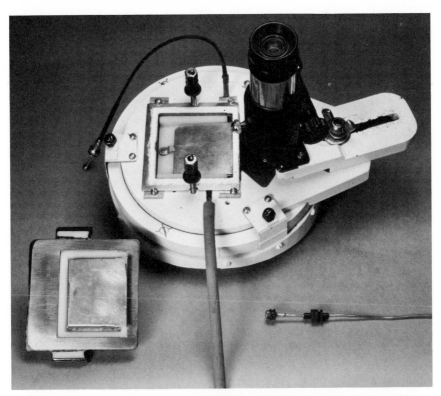

This picture of Evered Kreimer's guiding head shows the microscope in position for viewing the guide star. At the center of the head is the film holder, and at lower left in the picture is the box for containing the dry ice used to cool the film.

With the low-temperature camera cooling Tri-X film to $-109°$ Fahrenheit, 19th-magnitude stars could be recorded in eight to ten minutes. For many of the fainter Messier objects, a somewhat unconventional technique was employed: very heavy exposure followed by underdevelopment in low-contrast developer. It was thus possible to reach fainter than magnitude 19.5 in a dark sky if the seeing was good. This method was also useful for small subjects like planetaries, keeping film grain from showing too much in the enlargements.

When the Virgo group of galaxies was being photographed, bright sky conditions did not permit going much below magnitude 16. So, for three or four of these individual galaxies two negatives were superimposed during printing. With some bright objects, such as galactic clusters, the exposures were for only five minutes, and the negatives were developed to high contrast. This gave noticeably sharper images when exposures had to be made through turbulent air.

In almost all cases the negatives were printed on No. 5 paper, which helped bring out faint details, but at the cost of leaving heavily exposed areas a nearly featureless white. For some prints, such as of the Pleiades, it was desired to compress the emulsion latitude scale to bring out detail in bright areas. This was accomplished by preparing an out-of-focus positive transparency on sheet film, then using it as a dodging mask when making the print.

Of course, no one telescope can be best for photographing as varied a set of objects as those in the Messier catalogue. Many kinds of darkroom techniques are needed to bring out the desired details in all of them. However, in preparing this collection, the aim was not necessarily to present the "best possible" photograph, but one that could readily be compared with the visual impression.

Likewise, in photographing the drawings for reproduction, contrast and background density were kept at a level similar to that of the corresponding photograph. As a result, many of the drawings show a striking likeness to the photographs.

The Kreimer 12½-inch Cave reflector inside its shelter at Prescott, Arizona. Note the large slit and flip-off covering.

Our Messier Album

A S TOLD in the preceding chapter, the purpose of this album is to
bring to the amateur astronomer a homogeneous collection of visual
characterizations and photographs of the objects in Charles Messier's
famous catalogue. The descriptions are often accompanied by drawings,
which like the photographs are oriented with south up. The two excep-
tions are M31, where southeast has been placed at the top to preserve a
generous scale of reproduction, and M45, which has north up to aid binocular users. Detailed information about each photograph, including
exposure data and the scale of reproduction, is given in the table beginning on page 212.

Finder charts are also provided. They were drawn to show the location
of each Messier object in relation to a familiar or easily recognizable star
pattern. Each chart has north up, and the approximate scale is indicated
on it.

In the heading for each object is, where appropriate, its number in J.
L. E. Dreyer's *New General Catalogue* (1888), its *Index Catalogue* supplement (IC), or some other source. The cited right ascensions and declinations are for the epoch 1950. (Coordinates for the year 2000 are included
in the table beginning on page 212.)

M1

Basic data. The famous Crab nebula is an expanding cloud of gas from the explosion of a brilliant supernova observed in A.D. 1054 by Oriental astronomers. Modern measurements show an expansion rate of about 1,000 miles per second. Since Messier 1 is about 6,000 light-years distant, its present angular size of 6 by 4 minutes of arc corresponds to about 10 by 7 light-years.

The total light of the Crab nebula is equivalent to a star of visual magnitude 8 or 9. Near the center of the nebula is a 16th-magnitude star that is the collapsed core of the supernova. This object is the pulsar NP 0532, which every 0.033 second emits a pulse of radiation at radio, X-ray, and optical wavelengths. This pulsar is now believed to be a rotating neutron star, whose gradual loss of rotational momentum provides the energy by which the nebula shines.

The Crab's light is very strongly polarized in a varying pattern. To astrophysicists, this violently active nebula and its pulsar comprise one of the most fascinating objects in the sky. Most older handbooks mistakenly call it a planetary nebula.

NGC description. Very bright and large, extended along position angle approximately 135°; very gradually brightening a little toward the middle, mottled.

Visual appearance. Hand-held 7 x 50 binoculars show M1 as a dim patch, and it is easy in the 10 x 40 finder of the Mallas 4-inch refractor. Visually, a haze seems to surround the brighter middle, and the color appears slightly greenish.

On the best nights, an experienced observer may notice some streaks throughout the inner portion of the nebula, but they are extremely difficult to see. While all magnifications work well with the Crab nebula, medium powers are preferred for the 4-inch.

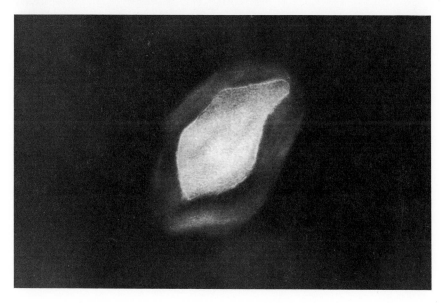

M2

Globular cluster in Aquarius

NGC 7089   21h 30m.9   −1° 03′

Basic data. Most of the stars of M2 lie in an area about eight minutes of arc in diameter, but long-exposure photographs show the cluster to be as large as 12 minutes. At a distance of 50,000 light-years, the latter diameter would correspond to a linear size of about 175 light-years. The total light of M2 is a match for a star of visual magnitude 7. It is quite easy to find this bright object in its star-poor field, as shown by the finder chart.

This cluster is rather poor in variable stars, about a score having been discovered. Most of these are RR Lyrae variables with periods of about a half day, and three are Cepheids with periods of 16 to 19 days.

But the brightest variable in M2 is one of the few RV Tauri stars known in globular clusters. It has alternating deep and shallow minima, and was discovered in 1897 by a skillful French ama-

teur, A. Chèvremont. The light fluctuations of the star from magnitude 12.5 to 14.0 "strikingly alter the appearance of the cluster," according to Helen Sawyer Hogg. The star lies on the eastern edge of M2, slightly north of center.

NGC description. Very remarkable globular cluster, bright, very large, gradually pretty much brighter toward the middle, well resolved into extremely faint stars.

Visual appearance. A beautiful object that is easily visible in binoculars. The Mallas 4-inch refractor does not resolve the cluster, except for a few bright members across the nebulous image. Visually, the most unusual feature is the dark curving lane that crosses the northeast corner of the cluster, as shown in the drawing. Though first thought by Mallas to be an illusion in the 4-inch refractor, this feature can also be detected in the photograph by Kreimer.

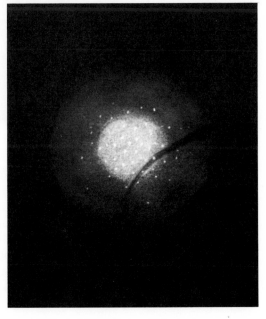

M3

Globular cluster in Canes Venatici

NGC 5272   13h 39m.9   + 28° 38′

Basic data. This tightly packed stellar swarm contains at least 44,500 stars, according to A. R. Sandage's counts on photographs made with the 200-inch telescope. The apparent diameter of a globular cluster depends on how it is observed, long-exposure photographs giving a much greater extent than visual scrutiny with small apertures. A compromise value for M3 is 10 minutes of arc. At the cluster's probable distance of 30,000 light-years, this corresponds to a linear diameter of about 90 light-years.

The total visual magnitude is about 6.4. M3 is approaching us at a rate of 150 kilometers per second. It contains about 170 faint RR Lyrae-type variable stars, more than any other globular cluster. These variables are used to determine the cluster's distance.

NGC description. A very remarkable globular; extremely bright and very large; toward the middle it brightens suddenly; it contains stars which are 11th magnitude and fainter.

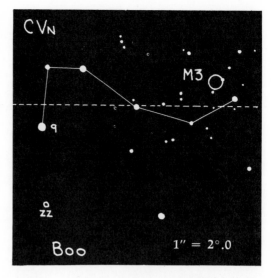

Visual appearance. A grand sight! In the 4-inch it consists of two concentric portions: a compact, very bright central area and a surrounding glow that fades uniformly outward to the edge. As the drawing indicates, a few outlying stars are resolved. The very grainy texture of the interior suggests that a larger aperture should show many stars.

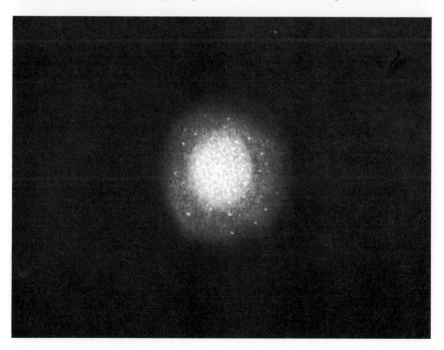

M 4
Globular cluster in Scorpius
NGC 6121   16h 20m.6   −26° 24′

Basic data. This conspicuous globular cluster near Antares was logged by Messier in 1764 but had been discovered in 1746 by L. de Chéseaux. Visually about 14 minutes in diameter, though appearing larger on photographs, M4 in total light is equivalent to a 6th-magnitude star. This great stellar swarm is about 10,000 light-years distant. Unlike some other globulars, such as M3 with hundreds of variable stars, M4 has only a few dozen of them.

NGC description. Cluster, with 8 or 10 bright stars in line . . . , readily resolved.

Visual appearance. A beautiful object. M4 can be seen with the naked eye on a clear dark night. In author Mallas' 4-inch refractor, this cluster is a well-defined circular glow with a brighter center. At 214x the outer parts of the globular are broken up into faint stars and the mid-portion appears partially resolved.

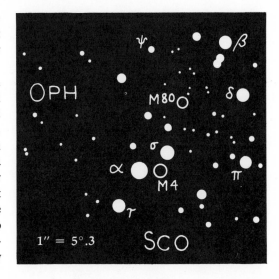

The interior of this globular presents the visual impression of a stubby band of stars, as indicated by the drawing and also in the Kreimer photograph. Inspection of the photograph reveals dark areas in the nuclear region and several curving spiral star chains in the outer portions of the cluster. These features were not noted visually in the 4-inch refractor.

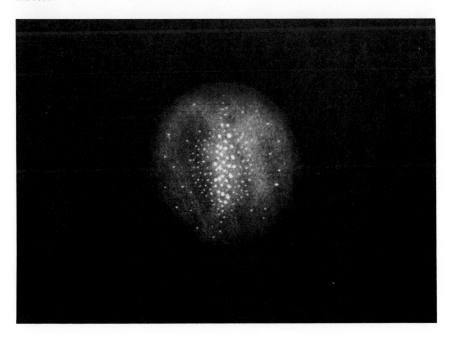

M5

NGC 5904   15h 16m.0   + 2° 16′

Basic data. One of the finest globular clusters in the heavens, Messier 5 was discovered in 1702 by Gottfried Kirch. Charles Messier himself did not come upon it until 62 years later. It is located to the northwest of the 5th-magnitude star 5 Serpentis.

The total light of M5 is equal to a 6th-magnitude star. Its angular diameter is 12 minutes of arc, which at the cluster's distance of about 30,000 light-years corresponds to 100 light-years.

This globular is noteworthy for its large number of variable stars. Most of the 100 or so now known are RR Lyrae stars with periods of approximately half a day. Only Omega Centauri and M3 are richer in variables.

The oval shape of M5 is evident in the Kreimer photograph shown above, and even more so on smaller-scale pictures. The elongation is from northeast to southwest.

NGC description. Very remarkable globular cluster, very bright, large, extremely compressed in the middle, stars from 11th to 15th magnitude.

Visual appearance. In the Mallas 4-inch refractor, M5 suggests a spider. One of its "legs" extends southward as far as the 5th-magnitude star 5 Serpentis, as shown in the drawing. This star is double, with a 10th-magnitude companion 11 seconds toward the northeast.

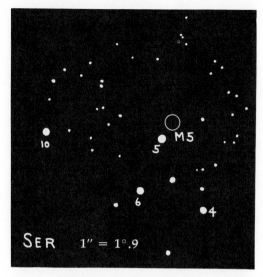

The core of this globular has a triangular shape, with a hint of partial resolution. M5 is a very beautiful sight at low to medium magnifications.

Admiral Smyth, who examined the cluster with a 6-inch refractor in 1838, wrote: "This superb object is a noble mass, refreshing to the senses after searching for fainter objects."

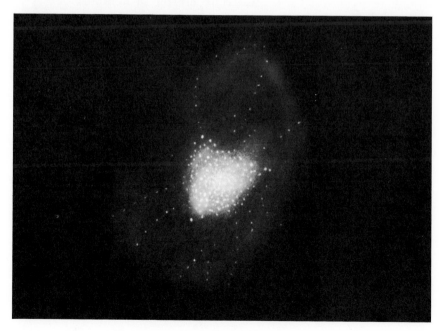

M6

Galactic cluster in Scorpius

NGC 6405   17h 36m.8   — 32° 11′

Basic data. This coarse open cluster, with a total light equivalent to a star of visual magnitude 4.3, is visible to the naked eye. It is only a bit fainter than M7, which lies about 4° to the southeast and is one of the brightest patches in the Milky Way. M6 is about 2,000 light-years distant and sprawls over an area approximately 26 minutes of arc in diameter.

Most of the brightest stars of M6 are blue or white, of spectral class *B* or *A*. Over 80 cluster members are brighter than 11th magnitude. The visually most conspicuous star, however, is decidedly reddish. It is BM Scorpii, a semiregular variable that changes very slowly between magnitudes 5½ and 7.

NGC description. Cluster, large, irregularly round, loosely compressed, stars from 7th to 10th magnitude.

Visual appearance. A pair of 7 x 50 binoculars resolves this cluster quite well, and for very small telescopes it is one of the finest sights in the heavens. In the Mallas 4-inch refractor it is a grand object with very little bunching of stars in the middle. The brightest stars delineate a trapezoidal figure. See finder chart on the facing page.

44

M7

Basic data. Almost one degree across, M7 contains 80 stars brighter than magnitude 10. Its total light is 3.5.

NGC description. Cluster, very bright, pretty rich, loosely ′ compressed, stars from 7th to 12th magnitude.

Visual appearance. Almost any optical aid will resolve this beautiful grouping. Many stars near its center appear yellow or orange. Overall, the cluster is circular in shape.

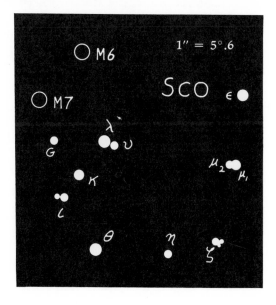

M8

Diffuse nebula in Sagittarius
NGC 6523   18h 01m.6   − 24° 20′

Basic data. Commonly known as the Lagoon nebula due to the great lane of obscuring matter that crosses its center, M8 is similar to M20, which lies only 1°.4 to the northwest. It is about 60 by 35 minutes of arc in size. The nebula may be 2,500 light-years distant, but this is uncertain.

The gases of this nebulosity are excited to shine by the 5.9-magnitude star 9 Sagittarii that lies within them. This star is even hotter than that which makes neighboring M20 glow, for its spectrum is classified as *O*5.

Adding to the view is the array of stars that make up NGC 6530. This fairly rich open cluster contains about 25 stars in an area 10 minutes of arc in diameter.

Not recognizable visually but clearly indicated in the photograph are many small, dark, compact spots silhouetted against the bright nebulosity. These are examples of the so-called globules — dense patches of interstellar dust that are believed to be contracting into stars.

NGC descriptions: M8. A magnificent object, very bright, extremely large and irregular in shape, with a large cluster.

NGC 6530. Cluster, bright, large, pretty rich, follows [lies east of] M8.

Visual appearance. The Lagoon nebula was the first deep-sky object observed by Mallas in the 1930's.

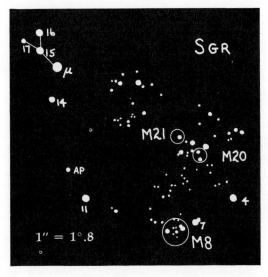

It is visible to the unaided eye in the rich Sagittarius Milky Way. The great extent of this nebula is revealed by 7 x 50 binoculars, and, when the air is not too steady, M8 seems suspended among the nearby stars. This illusion has never been noticed in the 4-inch.

M8 is one of the finest showpieces in the heavens. It is very complex and spangled with stars, though some regions are devoid of them. In the 4-inch at a power of 60, the nebula appears knotted and streaked by dark patches and rifts. Medium magnification shows many details. The largest dark lane is an easy feature for very small telescopes, while larger apertures reveal irregularities in its shape and brightness.

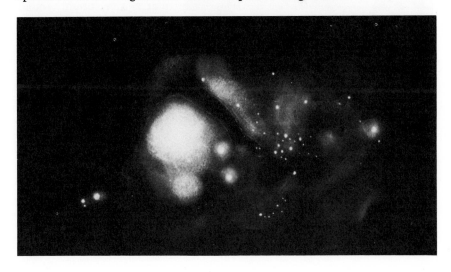

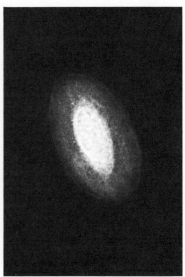

M9
Globular cluster in Ophiuchus
NGC 6333   17h 16m.2   −18° 28′

Basic data. Messier 9 appears much smaller than M5, being only about 3 minutes of arc across. Visually, its total magnitude is about 8½. Only about a dozen RR Lyrae stars are known in this globular. Wide-field photographs show that M9 is at the edge of a patch of dark nebulosity.

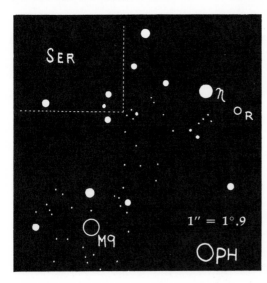

NGC description. Globular cluster, bright, round, extremely compressed middle, well resolved, stars of 14th magnitude.

Visual appearance. This cluster is impressive in the 4-inch at 120 power. The large, bright central region is oval, as is indicated in the photograph, which also shows the surrounding halo to be round. In surface texture M9 seems to be less grainy than M3 in Canes Venatici, which appears much larger.

M10

Globular cluster in Ophiuchus

NGC 6254   16h 54m.5   — 4° 02′

Basic data. This 7th-magnitude cluster is about 8 minutes of arc in diameter. It lies at a distance of perhaps 16,000 light-years and has a linear diameter of 80. Messier first saw M10 on May 29, 1764.

NGC description. Remarkable globular, bright, very large, round; gradually brightening to a much brighter middle; well resolved with stars 10th to 15th magnitude.

Visual appearance. A beautiful globular, this is one of the best for small apertures, with many stars visible in the Mallas 4-inch refractor. The central region appeared pear-shaped, with grainy texture at moments of steady seeing. At 120x, bright knots were noticed in the outer regions. A finder chart for M10 is included under M12.

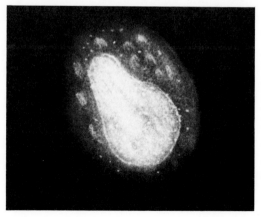

Galactic cluster in Scutum

M11

NGC 6705   18h 48m.4   −6° 20′

Basic data. One of the richest and most compact open clusters, M11 lies in the extremely rich Milky Way region of the Scutum star cloud. It was discovered in 1681 by the German observer Gottfried Kirch, who described it as "A small, obscure spot with a star shining through and rendering it more luminous."

Ake Wallenquist has estimated that M11 contains more than 600 members brighter than 15th magnitude. These are in an area about 12 minutes of arc across, which at the cluster's distance of 6,000 light-years corresponds to a linear diameter of 21 light-years.

NGC description. Remarkable cluster, very bright, large, irregularly round, rich, one star of 9th magnitude among stars of the 11th magnitude and fainter.

Visual appearance. One of the finest views in the heavens for small apertures, and even visible to the naked eye on dark and

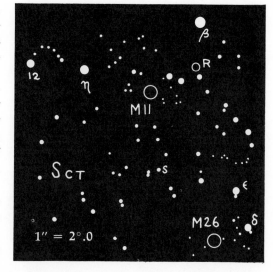

moonless nights. A 2-inch will just show a few of M11's brighter stars; while a 3-inch begins to resolve many others.

Visually, the cluster has a distinct shape, described by Admiral Smyth as somewhat resembling a flight of wild ducks. This impression is closely matched in the 4-inch refractor, which author Mallas used to make the drawing. Note the fan-shaped image with its dark arch. Two 8th-magnitude stars southeast of the cluster were easy in the 4-inch.

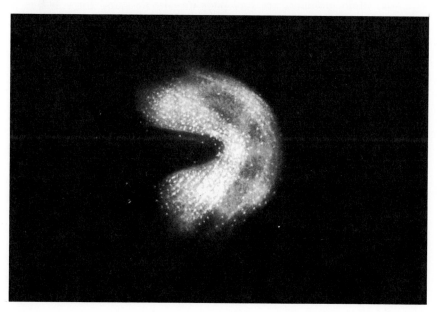

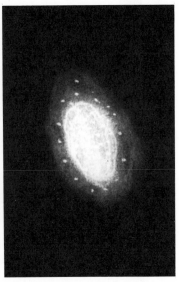

M12

Globular cluster in Ophiuchus

NGC 6218   16h 44m.6   −1° 52′

Basic data. M12 is nearly a twin of M10, having about the same total magnitude and apparent diameter, and lying at roughly the same distance from us. Both globulars are very poor in variable stars, for only one of these objects is known in M12 and three in M10.

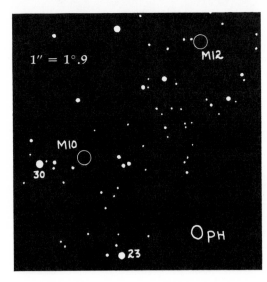

NGC description. Very remarkable globular, very bright and large, irregularly round; gradually much brighter toward the middle; well resolved, stars of 10th-magnitude and fainter.

Visual appearance. M12 is a fine object, though its stars appear to be very loosely concentrated. The brighter ones were resolved by the 4-inch at 120x. Little or no grainy texture was noticed in the central region, where the stars are much closer together.

M13

Globular cluster in Hercules

NGC 6205   16h 39m.9   +36° 33′

Basic data. M13 is one of the finest globular clusters in the heavens and a showpiece for Northern Hemisphere observers. With a total light equivalent to about a 6th-magnitude star, it can be seen with the unaided eye in a dark sky, as it was by Edmond Halley, who discovered this cluster in 1714, on the western side of the well-known Keystone in the constellation of Hercules.

At a distance of 25,000 light-years, M13 is among the nearest globulars. An angular diameter of 23 minutes of arc has been found from star counts on photographs, corresponding to a linear diameter of about 170 light-years. But this includes faint outlying parts not seen in amateur telescopes, which may show M13 only half that large.

This great swarm contains several hundred thousand stars, it is thought. Toward the center of a typical globular cluster, where the stars

53

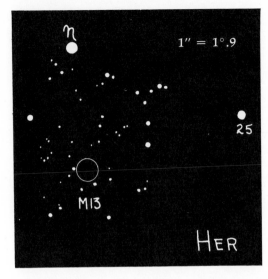

$1'' = 1°.9$

η

25

M13

HER

are most crowded, there are one or two per cubic light-year. Volume for volume, this is 500 times as many as in the solar neighborhood!

NGC description. Very remarkable globular cluster of stars, extremely bright, very rich, very gradually increasing to an extremely compressed middle, stars from 11th magnitude downward.

Visual appearance. M13 is a magnificent object in author John Mallas' 4-inch refractor, which will resolve some of its stars at magnifications of 120 and 250. As shown in the drawing, there are three or four apparently star-poor areas that can be detected by an experienced observer. A suggestion of these "lanes" also appears in the Kreimer photograph.

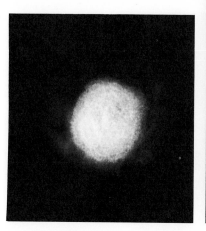

M14

Globular cluster in Ophiuchus
NGC 6402   17h 35m.0   −3° 13′

Basic data. This 8th-magnitude globular in a rich Milky Way field was discovered by Messier in 1764. It is considerably smaller than M10 or M12, its main body having a diameter of 3 minutes of arc.

This stellar swarm has a linear diameter of about 55 light-years (roughly the distance from the sun to Alpha Ophiuchi) and lies 23,000 light-years away. In contrast to M10 and M12, this globular contains over 70 known variable stars.

NGC description. Remarkable globular, bright, very large, round, extremely rich, very gradually becoming brighter toward its center, well resolved, 15th-magnitude stars.

Visual appearance. M12 has a nearly circular form in the 4-inch. The central two-thirds of the visual image is bright, but toward the outer edges the light fades rapidly. Some graininess was noticed at moments of steady seeing, giving the impression that a little more optical power would show some stars.

Globular cluster in Pegasus

M15   NGC 7078   21h 27m.6   +11° 57′

Basic data. Lying at a distance of roughly 40,000 light-years, Messier 15 has an apparent diameter of 12 minutes of arc (though visually it may appear only half as large) and shines at magnitude 7. M15 is the only known globular cluster with a planetary nebula. First reported by F. G. Pease in 1928, this 14th-magnitude planetary is only a second of arc in diameter.

Unlike the stars of M2, which are highly concentrated toward the cluster's center, those of M15 are more evenly distributed across the image. This is one of the richest globulars for variable stars, though far outclassed by Omega Centauri and M3. Nearly 100 variables are known in M15, most of them being RR Lyrae stars. (A laboratory exercise in astronomy by Owen Gingerich tells how M15 variables are used to determine the distance of the cluster; see the suggested reading list in this volume.)

NGC description. Remarkable cluster, very large and bright, irregularly round, very suddenly much brighter in the middle, well resolved into very small [faint] stars.

Visual appearance. The slightest optical aid reveals this grand globular. In the 4-inch, M15 appears circular, nestled in a fine star field. The center of the

cluster is very intense, with quick fading toward the edges, but M15 is not resolved by this aperture. Some of the faint extensions in the drawing are seen in the photograph, where the star at the field's north edge is 7 minutes of arc from the cluster's center.

M16
Cluster and nebula in Serpens
NGC 6611   18h 16m.0   −13° 48′

Basic data. The nebula is caused to shine by the cluster of hot blue and white stars. It is ½° in diameter and at a distance of 7,000 light-years from us. The finder chart is reproduced under M17.

According to M. F. Walker the cluster is only two million years old. Its brightest members, such as the conspicuous double star, are concentrated toward the northwest and are approximately 8th magnitude.

On the southern side of the nebula are several dark "elephant trunk" structures. Note also in the photograph two very dark round globules; one is amid the cluster stars, the other about 0°.1 to the south (⅔ inch above and slightly to the left).

The M16 star cluster was discovered by P. L. de Chéseaux in 1746, the same year he found M17. The surrounding nebulosity, however, seems to have been first noted by Charles Messier two decades later.

NGC description. Cluster, at least 100 bright and faint stars.

Visual description. Messier 16 is one of the most unusual objects in the sky and a fine sight at low power. The 4-inch reveals three nebulous regions and about 20 stars against an uneven background.

In this wide-field view of the summer Milky Way, taken by Duane Ruokonen of Milwaukee, Wisconsin, a tenth of all the Messier objects are seen, namely three nebulae (M8, M20, and M17), two globulars (M22 and M28), five open clusters (M16, M18, M21, M23, and M25), and the Milky Way patch, M24. Stars in the Teapot of Sagittarius are at bottom left and center, while Omicron Serpentis is at the right edge near the top of the field. Right of center and 1½ inches from the bottom is M8, above which are M20 and M21. M28 is near Lambda at the top of the Teapot, while M22 is above and left (northeast). About 3 inches from the bottom, left of center, is M25, and right of center is M23. Between them is the starcloud M24 (containing NGC 6603), and in succession above the cloud are M18, M17, and M16.

As a high school student, Mr. Ruokonen took the picture at Grindstone Lake, Wisconsin, on August 14-15, 1974. For the 20-minute guided exposure he used Tri-X film in a Pentax camera with a 55-mm. f/2 lens.

M 17

Nebula and cluster in Sagittarius

NGC 6618   18h 18m.0   −16° 12′

Basic data. This object takes its familiar name of Omega or Horseshoe nebula from its appearance in small telescopes. As in M16, hot stars excite the gas to shine. The main source of this excitation may be in a small cluster of stars which lies near the bright core of the nebula, according to Colin S. Gum.

Associated with the nebulosity is a cluster of about 35 stars scattered over an area 0°.4 across. This loose grouping is quite inconspicuous, however, and in a small telescope appears similar to a rich patch of the Milky Way itself.

Recorded in the photograph is a faint "V" of nebulosity that extends as much as half a degree to the east (right) from the curved portion of the Omega.

NGC description. Magnificent, bright, extremely large, extremely irregular shape, hooked like a "2."

Visual appearance. The Omega nebula dominates the field of the 4-inch, and only a few stars were noted in its vicinity. Though the dark areas in and around the Omega were easily seen, the faint patches shown in the drawing were difficult at 60 power.

The most conspicuous portion of the nebula is the straight bar, which appeared vividly white. The sky inside the hook looked particularly dark, which was perhaps a contrast effect.

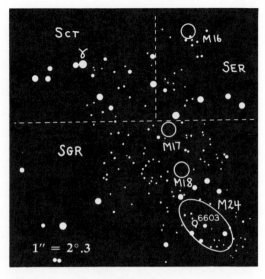

Messier 17 will challenge visual observers. Like the Orion nebula it repays careful and repeated study.

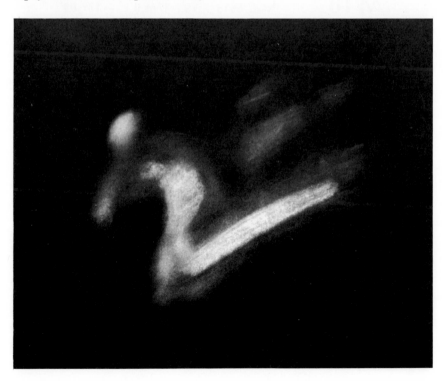

M18

Galactic cluster in Sagittarius

NGC 6613   18h 17m.0   −17° 09′

Basic data. The total visual brightness of M18 is roughly equivalent to a 7½-magnitude star. Containing only about 12 fairly bright stars in an area 0°.2 across, this open cluster is loose and poor. It lies at a distance of perhaps 6,000 light-years.

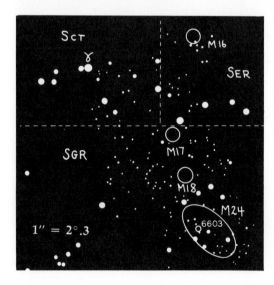

NGC description. Cluster, poor, very little compressed.

Visual appearance. Here is a pretty sight for very small telescopes, which may reveal more than a dozen stars. In the 4-inch refractor of author Mallas, the cluster seems to have a nebulous glow, which does not show in photographs and probably results from insufficient aperture for adequate resolution. Note how unevenly the field stars are distributed.

M 19

Globular cluster in Ophiuchus
NGC 6273   16h 59m.5   −26° 11'

Basic data. This 7th-magnitude cluster has an apparent diameter of 5 minutes of arc. At its estimated distance of 20,000 light-years, the linear diameter of M19 is about 30 light-years. Charles Messier first observed this globular on June 5, 1764.

Admiral W. H. Smyth, who examined M19 in his 6-inch refractor in 1837, saw it as a very compressed cluster of faint points of light, situated nearly midway between two telescopic stars.

NGC description. Globular, very bright, large, round, very compressed in the middle, well resolved. It consists of stars of 16th magnitude and fainter.

Visual appearance. This beautiful cluster appeared like a miniature Omega Centauri in the 4-inch. As seen from Covina, California, the central region had a grainy texture, as if it were on the threshold of resolution.

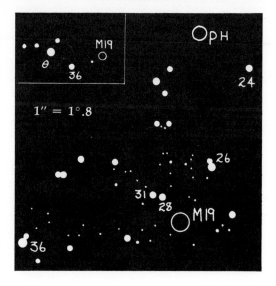

M 20

Diffuse nebula in Sagittarius

NGC 6514   17h 58m.9   −23° 02′

Basic data. This is the famous Trifid nebula, named for its three-lobed appearance. It is very large, with an extent of 29 by 27 minutes of arc, and lies at a distance of perhaps 2,200 light-years. Embedded in this giant cloud of cosmic gas and dust is a very hot 7th-magnitude star of spectral class $O7$ that causes the nebula to shine.

NGC description. A magnificent object, very large and bright, trifid, a double star involved.

Visual appearance. A very interesting nebula, easily visible in 7 x 50 binoculars in a beautiful Milky Way star field. The dark rifts that separate the lobes are readily seen in a 2.4-inch refractor, and a 4-inch reveals these lanes to be nonuniform in brightness. In the latter telescope, the northern section of M20 has a slight greenish hue; color pictures show it blue, for it is a reflection nebula whose dust is lit by neighboring stars.

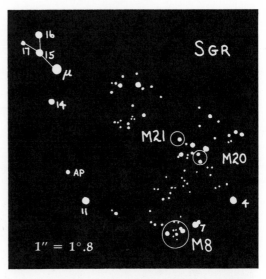

The "double star" mentioned in the *New General Catalogue* description lies near the central junction of the rifts. Actually this is the multiple system HN 40, discovered by William Herschel; its brightest component is the 7th-magnitude star noted above. The 4-inch shows three more companions.

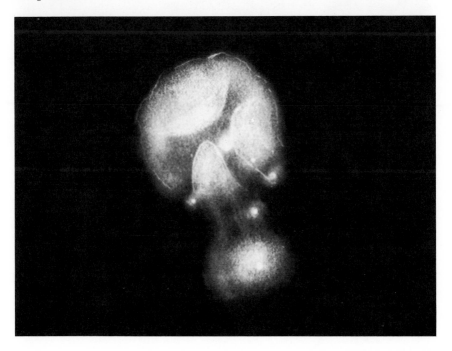

Galactic cluster in Sagittarius

M 21   NGC 6531   18h 01m.8   — 22° 30'

Basic data. The total light of this stellar assemblage is about that of a star of visual magnitude 6.5. According to the Swedish astronomer Ake Wallenquist, the cluster contains about 40 members brighter than magnitude 12 within an area 12 minutes of arc in diameter. One of several discordant published values of M21's distance is 3,000 light-years. This implies a cluster diameter of 10 light-years.

NGC description. Cluster, pretty rich, little compressed, stars from magnitude 9 to 12.

Visual appearance. Very fine and impressive. This rather small and compact cluster has a diamond shape in the 4-inch refractor. The photograph shows part of M20 at upper left. The finder chart is reproduced on the previous page.

Globular cluster in Sagittarius

M 22   NGC 6656   18h 33m.3   — 23° 58′

Basic data. The initial sighting of this cluster was probably by A. Ihle in 1665, making it the first globular known. It is a fine object, about 0°.3 in diameter and shining with a total light equivalent to a 6th-magnitude star. It lies 2½° to the northeast of the 3rd-magnitude star Lambda Sagittarii (the top of the Teapot).

M22 contains perhaps 70,000 stars. Of these, 25 are known variables, mostly of the RR Lyrae type with periods of roughly half a day. However,

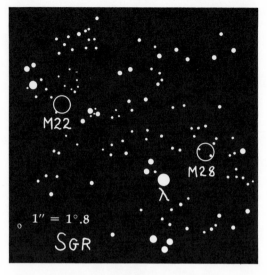

one star is a 200-day long-period variable; it is not a cluster member but merely situated along our line of sight.

M22 lies at a distance of about 10,000 light-years and has a linear diameter of roughly 50 light-years.

NGC description. Very remarkable globular, very bright and large, round, very rich and much compressed, stars from 11th to 15th magnitude.

Visual appearance. Visible to the naked eye, M22 is as impressive as M13 in Hercules. The 4-inch partially resolves the cluster, but not its center, which remains a solid glow. The two dark areas in the drawings are illusory, being outlined by strings of bright stars. These dark lines are noticed in telescopes with apertures as small as 2 inches.

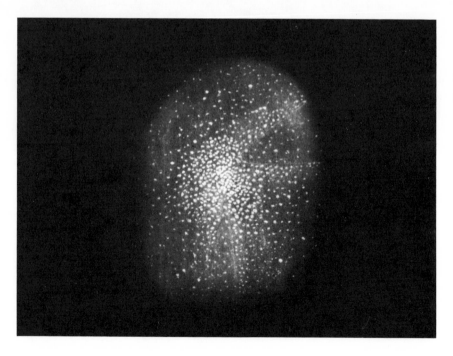

Galactic cluster in Sagittarius

M 23

NGC 6494   17h 54m.0   −19° 01′

Basic data. Lying at a distance of about 4,500 light-years, M23 has an apparent diameter of 25 minutes of arc, which corresponds to a linear size of about 30 light-years. The cluster is moderately rich (with about 120 stars) and has a total visual magnitude of 6.9.

NGC description. Cluster, bright, very large, pretty rich, little compressed, stars of 10th magnitude and fainter.

Visual appearance. A glorious sight in the 4-inch. The brightest stars in this irregularly shaped cluster form a pattern resembling a bat in flight. The photograph gives a hint of this likeness. M23 lies in a grand star field, most pleasing at low power.

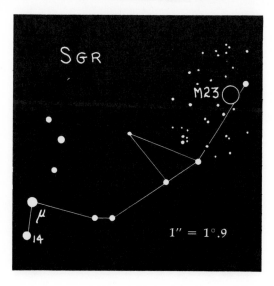

Milky Way patch in Sagittarius

M 24

NGC 6603   18h 15m.5   −18° 27′

Basic data. The designation M24 has been applied to two different objects, one within the other. Messier's description is of the patch of Milky Way, about 1° by 1½°, seen above in a photograph by Christos Papadopoulos of South Africa.

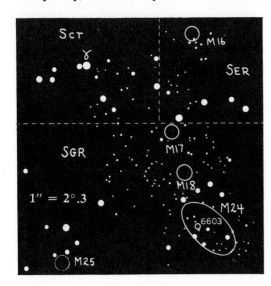

Near the center of the field is NGC 6603, shown in the photograph by author Kreimer on the facing page. Formerly, this was considered to be M24. The problem has been reviewed by K. Glyn Jones of the British Astronomical Association, who writes:

"Messier's own description for M24, which he said was 1½° in diameter, was 'a large nebulosity in which there are several

stars of different magnitudes.' In fact, M24 is not a true galactic cluster but a small detached portion of the Milky Way that has, in fact, an integrated magnitude of about 4½. It is quite easily seen with the naked eye under good conditions and is a fine sight in binoculars or a rich-field telescope.

"Within the boundaries of the M24 star cloud is a small denser area that Admiral Smyth called 'a gathering spot with much star dust.' This is the true galactic cluster NGC 6603, with a diameter of about 4½ minutes and an integrated magnitude of 11.4, according to Mrs. Hogg. . . . I have not been able to trace where the confusion concerning M24 began. In Dreyer's catalogue, NGC 6603 is equated with M24 and . . . in the appendix, a note taken from Sir John Herschel reads 'h 2004 = M24.'" M24.'"

NGC description. Remarkable cluster, very rich and very much compressed, round, stars of [12th] magnitude and fainter, in the Milky Way.

Visual appearance. In the 4-inch, NGC 6603 is a compact glow, containing stars forming a "V," as shown in the photograph. There are beautiful star fields in this area.

Galactic cluster in Sagittarius

M 25   IC 4725   18h 28m.8   —19° 17'

Basic data. This open cluster has an apparent diameter of 40 minutes of arc and shines with a total light equivalent to a 4.9-magnitude star, according to David F. Gray. Though many of its stars are conspicuous (the five brightest average magnitude 8.6), they are loosely scattered. Among stars brighter than magnitude 12.6, Ake Wallenquist found 86 probable cluster members. This system lies about 2,000 light-years from us, only a third as far away as its neighbor M18.

IC description. Cluster, pretty compressed.

Visual appearance. This is a superb sight for small apertures, with many colored stars. When the photograph is viewed with east at the top, the brightest stars of M25 form a straight-backed chair. As seen in the 4-inch, the cluster has about 50 stars. A finder chart is with M24.

M 26

Galactic cluster in Scutum

NGC 6694   18h 42m.5   −9° 27′

Basic data. Messier 26 is also situated in the Scutum star cloud, though it is less impressive than M11, which lies 3½° to the northeast. Altogether M26 contains about 90 cluster stars brighter than magnitude 15 in an area roughly 11 minutes of arc in diameter. As the distance of M26 is about 5,000 light-years, it is 16 light-years in linear diameter.

NGC description. Cluster, quite large, pretty rich and compressed, stars from 12th to 15th magnitude.

Visual appearance. The 4-inch refractor reveals more than 20 stars arranged roughly in the shape of a fan. In very clear skies, fainter stars are seen on an uneven background.

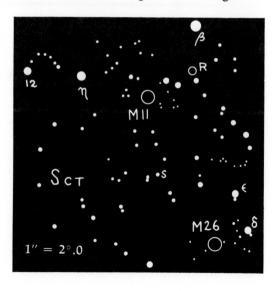

M 27

Planetary nebula in Vulpecula

NGC 6853   19h 57m.4   + 22° 35′

Basic data. Commonly known as the Dumbbell nebula, this famous planetary has an apparent size of 8-by-4 minutes of arc and shines at about 8th magnitude. Since M27 lies about 1,250 light-years away, the major axis of the "hourglass" is nearly three light-years long, about two thirds the distance from the sun to Alpha Centauri. At the center of the hourglass lies a 13th-magnitude star which has an estimated temperature of 85,000° Kelvin. Its radiation causes the rarefied gas to shine by fluorescence. As the finder chart shows, M27 lies not far to the south of the 5th-magnitude star 14 Vulpeculae.

NGC description. Magnificent object, very bright and large, binuclear, irregularly extended (Dumbbell).

Visual appearance. A superb planetary for low to medium magnification with small apertures; even the 10-power finder reveals details. Glowing quite greenish, M27 is one of the few planetaries to show vivid color in a small telescope.

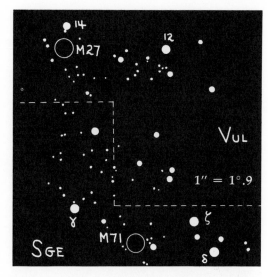

At low power, when the air is not too steady, the Dumbbell may seem three-dimensional and suspended in space, but this illusion is rare. While many stars are seen superimposed on the nebula, they are not included in the drawing. It is probably these and nearby field stars which are responsible for the "hanging in space" effect.

Seen as almost uniform in brightness, the disk of M27 is well defined, though the ends of the major axis are fuzzy and uneven. M27 is not the only dumbbell nebula, for M76 in Perseus, fainter and smaller, is similar in shape as seen in the 4-inch refractor.

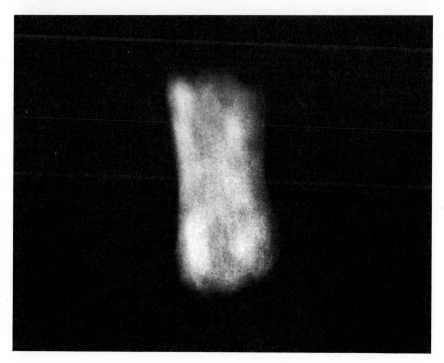

M 28

Globular cluster in Sagittarius

NGC 6626   18h 21m.5   −24° 54′

Basic data. Though Charles Messier discovered this globular in 1764, his relatively poor equipment revealed only an amorphous glow. It was not until William Herschel observed M28 that it was recognized as being a star cluster.

This 7th-magnitude globular has an apparent diameter of 15 minutes of arc, which at its distance of 15,000 light-years corresponds to a linear size of about 65 light-years. Like M22, its larger neighbor in Sagit-

tarius 2½° to the north-
east, M28 is relatively poor
in variable stars.

NGC description. Re-
markable globular, very
bright, large, round, in-
creasingly compressed in
the middle, well resolved,
stars from 14th to 16th
magnitude.

Visual appearance. In
the 4-inch this strangely
shaped globular somewhat
resembles a cucumber. The
inner and middle regions
of the Kreimer photograph
support this visual impres-

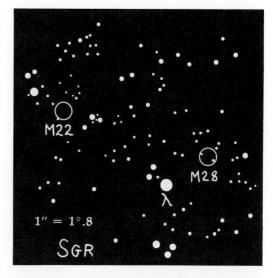

sion. However, only the central portion of the cluster could be seen well
in the refractor, not the faint outlying stars that form a roundish haze.

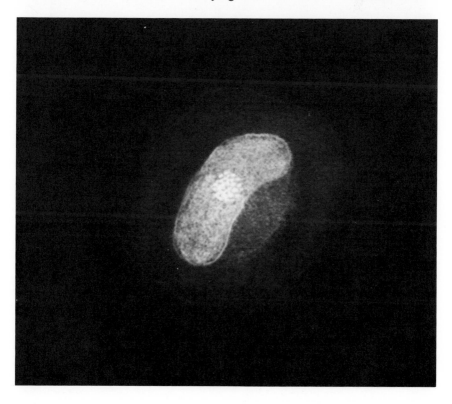

M 29

Galactic cluster in Cygnus

NGC 6913   20h 22m.2   + 38° 21′

Basic data. This coarse group of stars lies nearly 2° south of Gamma Cygni. Most authorities give the angular diameter of M29 as 6 or 7 minutes of arc, and its total light as 7th magnitude. About six stars are brighter than 9½. The cluster's distance is some 7,200 light-years.

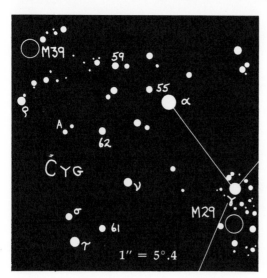

NGC description. Cluster, poor and little compressed, bright and faint stars.

Visual appearance. The brighter members of M29 form a stubby dipper, as seen in the photograph. The 10 x 40 finder gave an attractive view, but each increase in power reduced the cluster's beauty. With the 4-inch refractor, lowest power was best.

M 30

Globular cluster in Capricornus
NGC 7099   21h 37m.5   −23° 25′

Basic data. The apparent diameter of M30 is about 6 minutes of arc. At this globular cluster's distance of about 40,000 light-years, the corresponding linear diameter is 75 light-years. M30 has a total visual magnitude of 8.1 and is approaching us at 164 kilometers per second.

NGC description. Remarkable globular, bright, large, slightly oval. From its edge, it gradually brightens to a much more intense middle. Stars from 12th to 16th magnitude.

Visual appearance. A splendid object even in small apertures. The 4-inch refractor shows a bright, fuzzy central core, surrounded by a slightly fainter region having a sharp but irregular outline. Outside this is a vague glow with some individual stars. In all, the visual appearance of M30 is quite unusual for a globular cluster. If the photograph is viewed at arm's length (while squinting), these three zones are suggested.

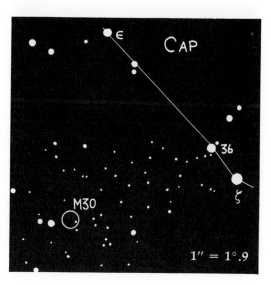

M31

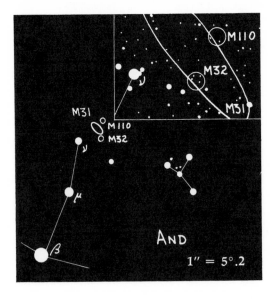

$1'' = 5°.2$

Basic data. M31 has a total visual magnitude of about 4; hence it can be seen with the naked eye. The finder chart shows how to "star hop" from Beta Andromedae (Mirach) to Mu and Nu and to the galaxy itself.

The longest axis of M31 is oriented roughly northeast-southwest in the sky, but the Kreimer three-part mosaic has been turned to fit the pages. Its overall length is more than 2°, but on the best photographs

with Schmidt telescopes at large observatories, M31 is some 3° in length.

Yet careful binocular estimates by the French astronomer Robert Jonckheere in 1952-53 gave a length of 5°.2 and a width of 1°.1. His 2-inch binoculars had all optical surfaces coated, with the oculars centered so that both eye pupils were fully used. Other precautions included 10 minutes of dark adaptation and exclusion of all light except from the galaxy's immediate vicinity.

The Kreimer hour-long exposures with the 12½-inch reflector show M31's spiral arms distinctly, even though the plane of the system is inclined only 15° to our line of sight. This object's spiral structure was not discovered visually, but the dark lanes of obscuring matter had been glimpsed over a century ago. The photograph, which has a scale of about one degree to 5½ inches, shows M31's small bright core, to the south (above and left) of which is M32, an elliptical galaxy separately listed by Messier.

The great Andromeda galaxy is of type Sb, marked by tightly wound spiral arms and a central condensation of intermediate size. M31 is approaching us at 68 kilometers per second. Like our Milky Way system, it is a member of the Local Group of about 20 galaxies. A recent determination of the distance to M31 is 2.2 million light-years, by

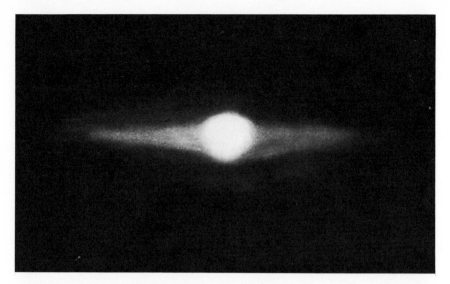

Henrietta Swope at Hale Observatories, in Pasadena, California. In the central condensation of M31 is a tiny, starlike nucleus, originally discovered visually with a 10½-inch refractor and about as bright as a 12th or 13th-magnitude star. Recent spectroscopic studies with the 120-inch Lick and 200-inch Hale telescopes show that it is a very dense swarm of stars, with a total mass of 160 million suns concentrated into a volume only 34 light-years in diameter. This nucleus is rapidly rotating, in a period of about 310,000 years.

Only 16 seconds of arc southeast of the nucleus is the spot where the 6th-magnitude supernova S Andromedae appeared in August, 1885, and faded from view the next year. The violence of that great stellar explosion is indicated by the fact that ordinary novae in M31 seldom become as bright as magnitude 16.

NGC description. Magnificent object, extremely bright, extremely large, very much extended.

Visual appearance. The Andromeda nebula is impressive in a small telescope, although the beginner may be disappointed that he cannot see its details. With the 4-inch refractor, author Mallas was not able to see the large extent of M31 that is photographed. The best way to detect the faint extensions shown in the drawing is to let the galaxy drift through the field of the telescope.

The brighter portions of the arms form a flattened diamond-shaped figure. The central condensation appears very intense and starlike at low power. The 4-inch at a magnification of 25 does not reveal a grainy central region, yet this visual characteristic was prominent to Mallas in other galaxies of the Messier catalogue.

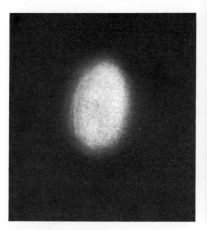

M 32

Galaxy in Andromeda

NGC 221   0h 40m.0   + 40° 36′

Basic data. M32 has a visual magnitude of 8.7 and is located 24 minutes of arc south of M31's core. It is one of a pair of bright companions to the Andromeda nebula, the other being 9th-magnitude NGC 205, which is now generally regarded as M110. M32 is an elliptical galaxy, with an apparent diameter of about 3 minutes of arc, which corresponds to a linear extent of 2,000 light-years.

NGC description. Remarkable, very bright, large, round, suddenly much brighter in the middle toward the nucleus.

Visual appearance. This beautiful galaxy is one of the best examples of an elliptical. Visually it has an oval form and bears magnification well. In the 4-inch, the brightness of M32 is pretty uniform nearly to the edge; then it fades rapidly into the sky background. In the lower left of the photograph of M32, the grayness is caused by outlying reaches of the Great Nebula, M31, as is also shown in the picture of that object.

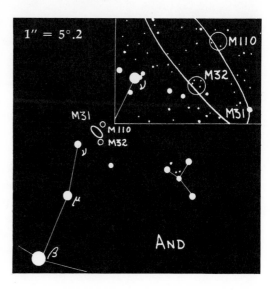

Galaxy in Triangulum

M 33   NGC 598   1h 31m.1   + 30° 24′

Basic data. M33 is a member of the Local Group of galaxies, which includes our Milky Way, the Magellanic Clouds, and M31. Because M33 has open, richly structured arms and a compact nucleus on photographs, it is classified as an Sc spiral. Its distance is about 2,300,000 light-years. The total light of this galaxy is equal to a star of visual magnitude 5.3. However, since M33 is about 1° across, its surface brightness is low.

NGC description. Remarkable, extremely bright and large, round, very much brighter in the middle to a nucleus.

Visual appearance. Author John Mallas has never been certain whether what he sees with the naked eye is M33 or the many faint foreground stars. Perhaps it is a combination of both. A pair of 7 x 50 binoculars shows M33 rather easily, but it is very faint and difficult in the 4-inch f/15 refractor. Instruments with smaller focal ratios will do much better, and low powers give best results. In the

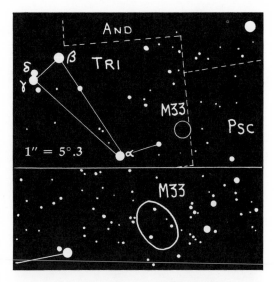

4-inch, M33 consists of a faint central region surrounded by fainter, soft gray patches, which are difficult to distinguish from nearby stars.

In field glasses or small short-focus instruments, author Mallas has sometimes seen M33 and the surrounding stars present a three-dimensional effect. It occurs when the air is not steady.

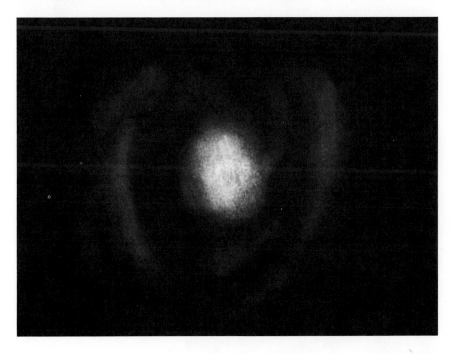

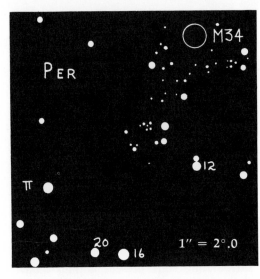

Basic data. First recorded by Charles Messier in 1764, this loose cluster of stars is just visible to the naked eye nearly midway from Algol to Gamma Andromedae. It lies 1,400 light-years from us according to H. L. Johnson, who measured a dozen stars in the group as outshining visual magnitude 9.0. However, the most brilliant, at 7.3, is a nonmember. M34 is about ½° in extent, but its brighter stars are concentrated in a smaller area.

NGC description. Bright, very large cluster, little compressed, of scattered 9th-magnitude stars.

Visual appearance. A very fine cluster which 7 x 50 binoculars will resolve quite well. It is so extended that a 4-inch may give a more attractive view than a larger telescope would. The main stars form a distorted X. The cluster's second brightest star, magnitude 7.9, is a close visual double with somewhat unequal components 1.4 seconds of arc apart. Known as Otto Struve 44, it was discovered by him about 1840 with a 15-inch refractor.

M34 is somewhat more than 2½° due north of the 5th-magnitude star 12 Persei.

M 35

Galactic cluster in Gemini

NGC 2168   6h 05m.7   + 24° 20′

Basic data. According to David F. Gray, the total visual magnitude of this cluster is 5.3, but this seems slightly too bright to Mallas. Within a diameter of 30 minutes of arc, Ake Wallenquist found about 120 cluster members brighter than photographic magnitude 13. The distance to M35 has been reliably determined as 2,800 light-years.

Just southwest of M35 is the faint compact cluster NGC 2158, at right ascension 6h 04m.3, declination + 24°.1 (1950 coordinates). Though conspicuous at upper left in the photograph, this 12th-magnitude object is difficult in the 4-inch, appearing as a mere patch. NGC 2158 is 4 minutes of arc in diameter.

NGC description. Cluster, very large, considerably rich, pretty compressed, stars from 9th to 16th magnitude.

Visual appearance. M35 is just visible to the naked eye, and even

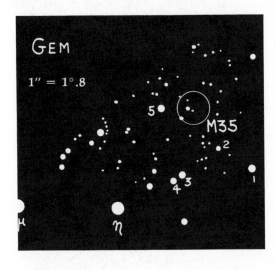

GEM

1″ = 1°.8

5

M35

2

3
4

1

μ   η

binoculars will reveal the brightest stars. In the 4-inch refractor at 25 power M35 is a splendid sight.

The cluster is nearly circular and the stars are quite uniformly distributed, with little concentration toward the center. Some other visual observers have mentioned patterns of stars in and around M35, but these were not seen in the 4-inch. M35 lies ½° west of the star 5 Geminorum.

M 36

Galactic cluster in Auriga

NGC 1960   5h 32m.0   + 34° 07′

Basic data. With an apparent diameter of 12 minutes of arc, M36 lies at a distance of 4,100 light-years and has a linear diameter of 14 light-years. The cluster contains about 60 stars between 9th and 14th magnitude, their total light being equivalent to a 6.3-magnitude star. M36 is fairly rich.

NGC description. Cluster, bright, very large and rich, little compressed, with scattered 9th- to 11th-magnitude stars.

Visual appearance. A grand view at low power, with marvelous color contrasts among the stars. In the 4-inch refractor there was a concentration of stars toward the center. Outward streamers of faint stars gave a crablike appearance. The cluster stars have a considerable range in magnitude. The finder chart is on the facing page.

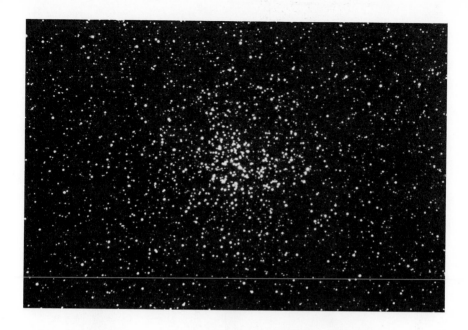

M 37

Galactic cluster in Auriga

NGC 2099   5h 49m.0   + 32° 33'

Basic data. M37, with an apparent diameter of 20 minutes of arc and lying about 4,600 light-years from us, has a linear diameter roughly twice that of M36. Containing about 150 stars to magnitude 12½, M37 has a total visual magnitude of approximately 6.2.

NGC description. Cluster, rich, pretty compressed in the middle, with large and small [bright and faint] stars.

Visual appearance. This is one of the finest open clusters in the heavens, lying in a very rich field of faint stars. Visually in the 4-inch at 60x, the cluster has a very elliptical outline, with bunches of bright stars at the ends; there appear to be more than 150 stars.

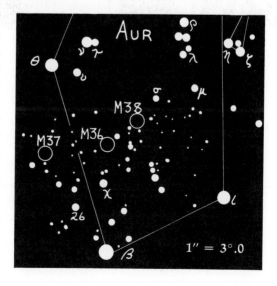

M 38

Galactic cluster in Auriga

NGC 1912   5h 25m.3   + 35° 48′

Basic data. The apparent diameter of this cluster is the same as that of M37. Since M38 is nearer, some 4,200 light-years away, its actual size must be less, about 21 light-years. M38 contains some 100 stars. The total visual magnitude, according to D. F. Gray, becomes 6.2 if we eliminate the 6th-magnitude star ½° south of the cluster.

NGC description. Cluster, bright, very large and rich, with an irregular figure, large and small stars.

Visual appearance. For small apertures, this is a beautiful cluster in a splendid field. Mallas disagrees with T. W. Webb's classic description of it as a "noble cluster arranged as an oblique cross: pair of larger stars in each arm." Instead, the 4-inch refractor showed M38 as square-shaped, with a clump of stars at each corner. The central star mentioned by Webb is not particularly conspicuous in the 4-inch. The finder chart is given under M37.

Galactic cluster in Cygnus

M 39   NGC 7092   21h 30m.4   + 48° 13'

Basic data. This bright open cluster is located nearly 10° east and a little to the north of Deneb. Its angular diameter is 30 minutes of arc. D. F. Gray has determined the total magnitude of M39 in visual light as 6.0, by summing the photoelectrically measured brightnesses of its individual stars. Visual estimates of the total magnitude by several observers average 5.4, however. M39 is one of the nearest open clusters, only 900 light-years from us according to H. L. Johnson. Turn the page for the finder chart.

NGC description. Very large, very poor cluster, very little compressed,

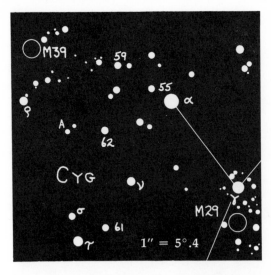

of 7th- to 10th-magnitude stars.

Visual appearance. M39 forms an equilateral triangle, with a bright star in each corner, outlining some 25 stars brighter than magnitude 10½.

This cluster is visible to the naked eye on dark, moonless nights. Resolvable in 7 x 50 binoculars, it is a pretty sight in very small telescopes. M39 is less impressive in the limited field of the 4-inch.

	Double star in Ursa Major
M 40	Winnecke 4 12h 20m.0 + 58° 22′

Basic data. In searching for a nebula said by the 17th-century observer Johann Hevelius to exist in this vicinity, Messier could find only a pair of faint stars, to which he nevertheless gave a number in his catalogue. In comparing Messier's description with the sky, author Mallas noted the

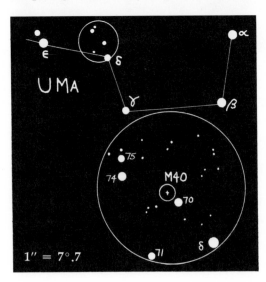

double star Winnecke 4 at the right position. It had been reobserved at Pulkovo Observatory in 1863. The two components are of visual magnitudes 9.0 and 9.3, and their separation on the sky is 49 seconds of arc.

Visual appearance. The double was very easy with the 4-inch refractor at 25x. No nebulosity was noted around the pair, which is widely split in the photograph opposite. It shows the 13th-magnitude galaxy

NGC 4290 forming a right triangle with 6th-magnitude 70 Ursae Majoris and M40. This barred spiral, about 2 by 1 minutes in size, could not be seen in the 4-inch.

Owen Gingerich has provided the following translation of Messier's original description, from *Mémoires de l'Académie Royale des Sciences,* 1771: "The same night of October 24-25, [1764] I searched for the nebula above the tail of the Great Bear, which is indicated in the book *Figure of the Stars,* second edition. Its position in 1660 was right ascension 183° 32' 41", declination +60° 20' 33". By means of this position, I found two stars very near each other and of equal brightness, about 9th magnitude, placed at the beginning of the tail of the Great Bear. One can hardly distinguish them in an ordinary [nonachromatic] refractor of 6 feet [length]. Their position is 182° 45' 30", +59° 23' 50". We presume that Hevelius mistook these two stars for a nebula."

When precessed to 1950, Messier's position agrees almost exactly with that of Winnecke 4, but Hevelius' place shows that actually he observed the 5th-magnitude star 74 Ursae Majoris, more than a degree away.

M 41

Galactic cluster in Canis Major

NGC 2287   6h 44m.9   − 20° 42′

Basic data. About 4° south of Sirius, this bright stellar grouping has a 6th-magnitude star (12 Canis Majoris) near its southeastern edge. Though authorities disagree, the total visual magnitude is approximately 5.2.

Messier 41 is about 2,400 light-years distant from us, according to a thorough study published in 1954 by A. N. Cox. He gives the angular diameter as about 20 light-years. The famous red star at the center of this coarse cluster has a visual magnitude of 6.9 and a *K*3 spectrum.

Long ago, T. H. Espin suggested that this star may be variable.

NGC description. Cluster, very large, bright, little compressed, stars of 8th magnitude and fainter. [The *New General Catalogue,* although correctly stating that this cluster was observed by Flamsteed and Le Gentil before Messier, incorrectly calls it M14.]

Visual appearance. A grand view in the Mallas 4-inch refractor, and indeed one of the finest open

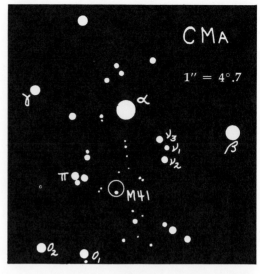

clusters for very small apertures. The brighter members form a butterfly pattern, but the cluster as a whole is circular, with little concentration. The 4-inch shows the Espin star as plainly reddish.

M42
M43

NGC 1976   5h 32m.9   − 5° 25′
Diffuse nebulae in Orion
NGC 1982   5h 33m.1   − 5° 18′

Basic data. To the naked eye M42 appears as a haze surrounding the 4th-magnitude star Theta Orionis, in the Sword of the Hunter. In small telescopes it is about a degree in diameter, overlapping its smaller neighbor to the north, M43. Actually, they are the brightest portions of a vast sea of faint nebulosities that photography has shown to cover almost the entire constellation.

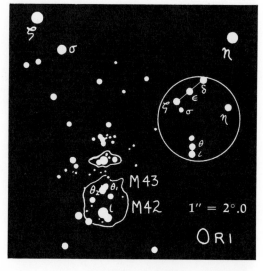

The Orion nebula was seen as early as 1610 by Nicholas Peiresc and in 1618 by J. B. Cysatus. Christiaan Huygens studied it carefully in 1656. Messier 43, however, was not recognized until many years later, appearing on a drawing made by the French scientist J. J. Mairan in 1731.

M42 is one of the brightest examples of an ionized-hydrogen (H II) region surrounding hot, young stars, such as those in the multiple system Theta[1] Orionis. Its four brightest members form the Trapezium.

The age of the central part of the Orion nebula seems to be roughly 30,000 years, about the same as the brighter Trapezium stars. It is one of the youngest nebulae in the sky. The distance to this glowing gas cloud is still somewhat uncertain, but is on the order of 1,000 light-years.

In the nebula and its surroundings are more than 500 known variable stars. The great majority are fluctuating irregularly in brightness within narrow limits. Many faint flare stars are seen.

NGC description. M42: Magnificent, Theta Orionis and the Great Nebula. M43: Remarkable, very bright and large, round with a tail, much brighter in the middle, contains a star of magnitude 8 or 9.

Visual appearance. Here is one of the most remarkable areas in the heavens. So many details are visible in even a small telescope that it is difficult to make a realistic drawing. The uneven surface brightness, fine filaments, and mottling near Theta Orionis are very hard to depict. With

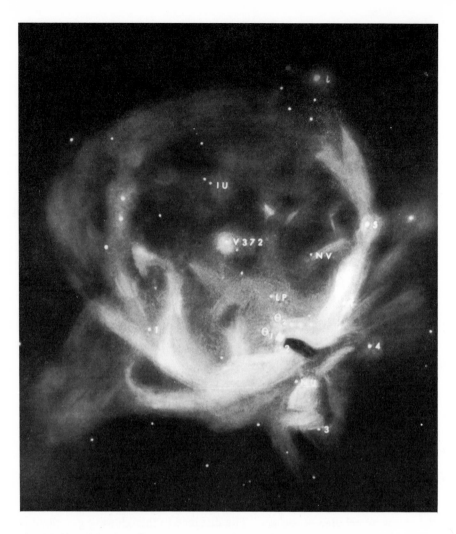

the Mallas 4-inch refractor, powers from 25 to 60 give the best general views of the nebula, whose greenish glow fills the entire field. A very conspicuous dark wedge (called the "fish's mouth" by Admiral Smyth) intrudes from the northeast, with Theta[1] Orionis near its tip.

From both sides of the wedge, great luminous bands curve away, forming a ring that can be traced through nearly its full circumference under favorable conditions. Lying on the eastern part of this ring is T Orionis, the prototype of the nebular variable stars. Just north of the wedge is the roughly triangular detached bright patch designated as M43. It is crossed by a dusky streak which narrows toward the east.

Other labeled variables are IU, LP, NV, and V372 Orionis; key features are numbered 1 to 5. Near the star 1, the ring splits into three branches, of which the northern and middle are faint. The third and brightest branch extends southward inside the ring, then veers sharply westward to end in the "hammerhead." Near the center of the ring lies the 8th-magnitude variable star V372 Orionis, which seems very green in the 4-inch and fuzzy during moments of good seeing. North and east of V372 are light and dark patches, a contrast of greens and grays.

Do not expect to see all the features of the Orion nebula on a first inspection. With favorable sky conditions, growing experience will reveal many delicate contrasts.

The Trapezium region should also be viewed with high powers. With 200x and 250x on the 4-inch, the wedge appears faintly luminous and its edge twisted. Below is a map of the Trapezium drawn 90 years ago by S. W. Burnham, the famous double star observer, to show the stars visible in the 36-inch refractor of Lick Observatory. The Trapezium consists of stars *A, B, C,* and *D,* which are labeled not in order of decreasing brightness but from west to east. *A* was discovered in 1975 to be an eclipsing variable that fades by a full magnitude for about 20 hours every 65.4 days; at minimum light *A* matches *D* in brightness. *B* is the variable star BM Orionis, also an eclipsing binary, with a period of 6.5 days; it is always the faintest of the four Trapezium stars.

Stars *E* and *F* can be seen in large amateur telescopes, but *G* and especially the double *H* are far too difficult. The cross marks an object glimpsed at the threshold of the 36-inch refractor by the eagle-eyed E. E. Barnard on two nights in 1888-89, but it is so excessively faint that it has never been distinctly seen by anyone else.

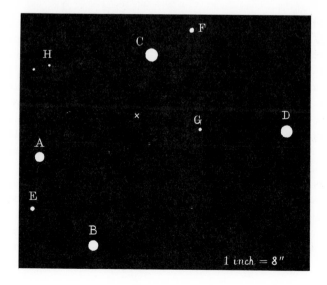

1 inch = 8 ''

Galactic cluster in Cancer
M44
NGC 2632   8h 37m.5   +19° 52'

Basic data. Easily visible to the naked eye as a fuzzy glow 1° across and about as bright as a star of magnitude 4.5, the Praesepe or Beehive was known even in ancient times as a nebulous object. Galileo, on viewing it in his telescope, wrote in 1610: "the nebula called Praesepe, which is not one star only, but a mass of more than 40 small stars."

This is still a good description of its appearance in very small telescopes, but about 350 stars brighter than magnitude 17 are known in the cluster area. Among them, but not a member of M44, is the Algol-type variable star S Cancri, which varies between visual magnitudes 8 and 10 in a period of 9.48 days.

M44 lies at a distance of about 500 light-years, being one of the nearest open clusters. The central region of bright stars is about 13 light-years across, but some outlying members increase the cluster's total extent to about 40 light-years. The member stars are moving through space together, sharing an annual proper motion of 0.037 second of arc southwestward and a recessional velocity of 33 kilometers per second.

Praesepe is a Latin word meaning "manger." In keeping with this theme the two stars flanking the cluster, Gamma and Delta, have Latin names meaning northern and southern ass. Both stars are conspicuous in the photograph below, taken by William Henry, which has north up, matching the orientation of the finder chart, but inverted with respect to the large-scale photograph on the facing page.

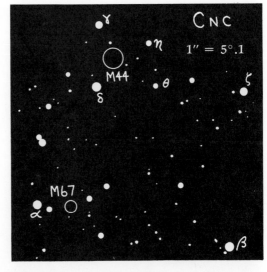

NGC description. None.

Visual appearance. Small binoculars resolve the Praesepe cluster into many stars for people with good eyesight. Excellent views are provided by 7 x 50 binoculars and rich-field telescopes. While M44 is too sprawling a group to give a good view in the 4-inch even at 25x, this instrument does show well the colors of the brighter stars.

Galactic cluster in Taurus

M45

3h 43m.9 + 23° 58'

Basic data. The Pleiades cluster has special fascination for amateur and professional alike. Otto Struve wrote in 1951: "It has been estimated that more photographs of the Pleiades have been taken by astronomers than of any other object in the stellar universe. Similarly, it is no exaggeration to say that more refined observations of all kinds have been made of the Pleiades than any other distant object."

A rich folklore has grown up about this conspicuous naked-eye star group. Our common name, the Seven Sisters, expresses the approximate number of stars visible to the unaided normal eye. (Patrick Moore once asked British television viewers to report the number of Pleiads they could see without optical aid; 73 percent of the replies said six, seven, or eight stars.)

Telescopically, this cluster contains several hundred members, most of which lie within 1° of the brightest, 2.9-magnitude Alcyone. Because all these stars must be at nearly the same distance from us (about 400 light-years), the brightest ones are also intrinsically the most luminous. They are all hot blue stars of spectral class *B*.

The total light of the Pleiades is equivalent to a star of visual magnitude 1.38, according to David F. Gray, who summed up the light contributed by the individual members. A nearsighted observer, who by removing his glasses can see bright stars as large disks, may confirm this value roughly by a comparison with Alpha Persei, magnitude 1.8.

It is fairly easy for amateurs to photograph traces of the wispy nebulosity in which this cluster is embedded, as shown by the Kreimer photograph. Stars as faint as magnitude 16 are recorded in a 10-minute exposure with a cooled emulsion. North is up, to aid comparison with the star charts of this cluster. At the left edge is Alcyone, while Taygeta is in the top right corner. In making the print, a dodging mask was used to avoid overexposing the nebulosity. Because of its low surface brightness, the nebulosity is hard to observe visually, but the least difficult part to glimpse extends southward from Merope. Also known as IC 349, the Merope nebula was discovered in 1859 by W. Tempel with a 4-inch

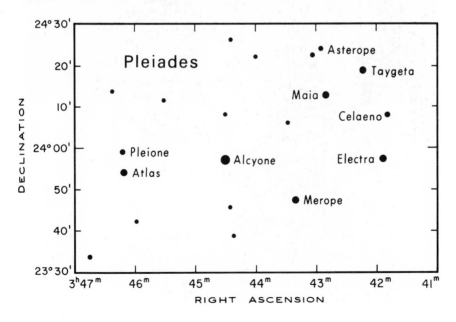

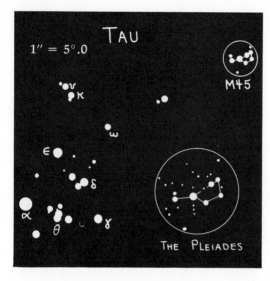

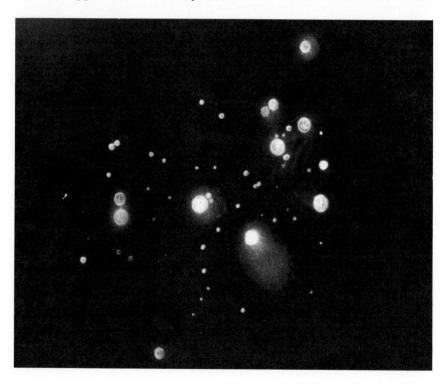

refractor. He described it as a faint stain. To view it requires observing skill and good skies rather than a large telescope.

The Pleiades nebulosity is a dust cloud faintly visible by reflected starlight. This was first shown in 1912 by E. C. Slipher at Lowell Observatory, who exposed a spectrogram for 21 hours to find that the nebula's spectrum matched that of the brighter Pleiads.

NGC description. Curiously, Messier 45 is not listed in the *New General Catalogue,* which also omits the Hyades and the Coma Berenices star cluster.

Visual appearance. In his youth, author Mallas could see a dozen

Pleiads with his unaided eye. The cluster is glorious in 7 x 50 binoculars, and in a 2-inch refractor at 15x is one of the finest sights in the heavens.

With larger telescopes, the Pleiades cluster is less striking, since only part of it can be viewed at once. In a 4-inch refractor at 60x the sight may be disappointing.

For many years Mallas searched for the Merope nebula, but without success. However, in the clear sky of Arizona, the 4-inch revealed a misty oval patch with Merope in its northern part. Normally, even on a clear dark night a 6-inch is needed to see IC 349.

More surprising were observations of the nebulosity near the stars Maia, Taygeta, and Celaeno. In the 4-inch, fine streaks were seen crossing an irregularly shaped haze, similar to the appearance in the Kreimer photograph. Nebulosity was also detected around Alcyone, but it was very weak and ill-defined.

The relation of the Pleiades to the familiar V-shaped Hyades is shown in the photograph above, taken by Walter J. Semerau. In the sky these two clusters are separated by about 10°.

M 46

Galactic cluster in Puppis

NGC 2437   7h 39m.6   −14° 42′

Basic data. This fairly rich open cluster, lying at a distance of about 5,400 light-years, contains some 200 members brighter than 14th magnitude within an area 28 minutes of arc in diameter. The total brightness of M46 is equivalent to a 6th-magnitude star.

As shown in the photograph, the planetary nebula NGC 2438 seems to lie near the northern fringe of the cluster. However, it is likely that this planetary is not part of M46 but a superimposed foreground object.

There is also a planetary inside the Pegasus globular cluster M15, but in that instance the two objects appear to be associated.

NGC description. Remarkable cluster, very rich, bright, and large, involving a planetary nebula.

Visual appearance. A magnificent cluster at low power. It is circular without any pronounced bunching of stars near the center, yet the brightest ones form many geometrical patterns. The planetary nebula is difficult to locate. It is oval, nonuniform in brightness, and blue-gray in color. In the 4-inch refractor of author Mallas, NGC 2438 is best seen at medium to high power.

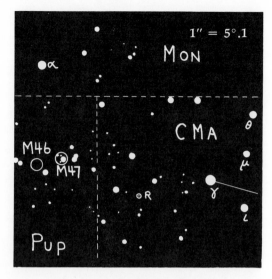

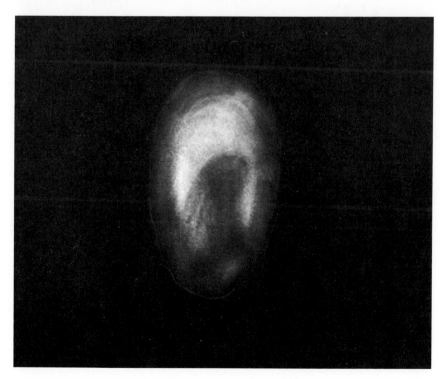

Galactic cluster in Puppis

M 47
NGC 2422　　7h 34m.3　　−14° 22′

Basic data. With a total magnitude of 4.6, M47 is visible to the unaided eye. It contains about 50 stars in an area 25 minutes of arc in diameter, the brightest one being of visual magnitude 5.7. About 1,600 light-years from us, M47 has a linear diameter of 12 light-years.

For many years it was regarded as one of the "missing" Messier objects, for no cluster could be found in the position recorded by Messier. However, in 1934 Oswald Thomas identified M47 as NGC 2422, and in 1959 T. F. Morris suggested that Messier had made an error in signs in entering the position. M47 now has two numbers: 2422 from William Herschel's discovery in 1785 and 2478 from Messier's discovery in 1771.

NGC description. Cluster, bright, very large, pretty rich, with large and small [bright and faint] stars.

Visual appearance. A beautiful coarse open cluster of bright stars, lying in the heart of the rich Puppis Milky Way. With the 4-inch, many colored stars were seen at low powers. The finder chart is on the preceding page.

M47 contains the fine double star Σ1121, located near the center of the Kreimer photograph (note the double diffraction spike). The components are both magnitude 7.9, separated by 7.4 seconds of arc in position angle 305°.

M 48

Galactic cluster in Hydra

NGC 2548   8h 11m.2   −5° 38′

Basic data. Also once a "missing" Messier object, M48 is now considered to be NGC 2548, for according to Owen Gingerich Messier's published position was five degrees in error. About 30 minutes in diameter, this open cluster has threescore members brighter than 13th magnitude. Its total light is magnitude 5.8, and its distance from us possibly 1,500 light-years.

NGC description. Cluster, very large, pretty rich, pretty much compressed toward the middle, 9th to 13th magnitude stars.

Visual appearance. A superb object in the 4-inch refractor and even partly resolved in binoculars. At 60 power, M48 is nearly circular and the brightest stars appear concentrated toward its center.

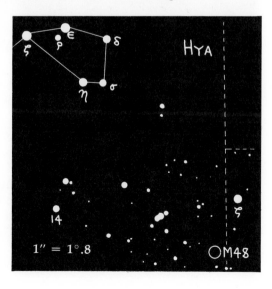

Galaxy in Virgo

M49

NGC 4472   12h 27m.3   +8° 16′

Basic data. This object was discovered in 1771 by 19-year-old Barnabas Oriani, who later became a distinguished Italian astronomer. It has an integrated magnitude between 8 and 9 and is 3 by 2 minutes in angular size. About 10 minutes to the south is the very faint peculiar galaxy NGC 4470, shown at the top in the photograph.

M49 is an E4 elliptical galaxy, lying about 70,000,000 light-years from us, the same distance as M61 and M104. The spectrum of M49 shows a red shift of 855 kilometers per second.

Photographs taken with the largest telescopes reveal many faint globular clusters surrounding M49.

About 1830 the star just east of M49 was observed several times by Sir John Herschel with an 18¾-inch reflector. He described it as 13th magnitude and situated outside the edge of the nebula. M49 itself was characterized by him as being very bright, round, and 2 minutes of arc across.

NGC description. Very bright, large, round, much brighter toward the middle, mottled.

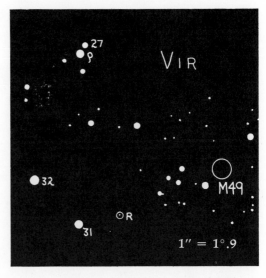

Visual appearance. In the 4-inch refractor, M49 resembles a globular cluster or the head of a comet, but its bright central region is more sharply defined than usual among galaxies. Both the photograph and the drawing show a broad, faint halo.

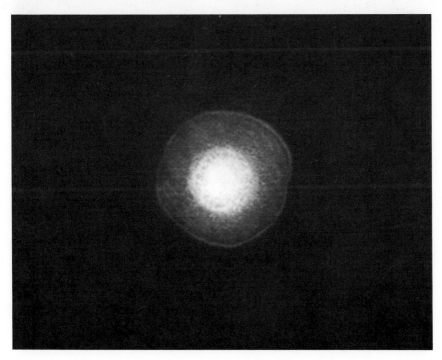

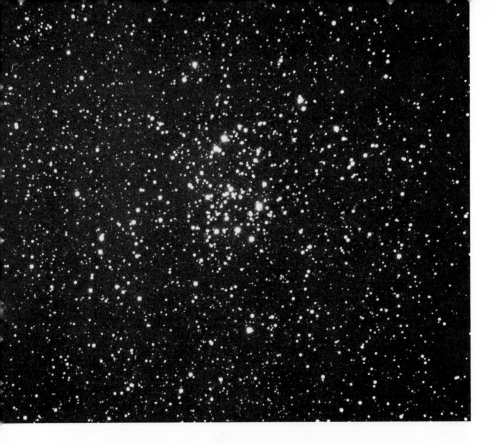

M 50

Galactic cluster in Monoceros
NGC 2323 7h 00m.5 − 8° 16′

Basic data. Charles Messier discovered this cluster on April 5, 1772, while observing the comet of that year. Kenneth Glyn Jones points out that M50 may have been seen by G. D. Cassini before 1711.

About 50 cluster members are found inside a circle with half the diameter of M48. Lying about 3,000 light-years away, M50 has a linear diameter of 13 light-years.

NGC description. Remarkable cluster, very large, rich, pretty compressed, elongated. The stars range from 12th to 16th magnitude.

Visual appearance. This grand cluster is well defined and circular in the 4-inch refractor; it could be resolved in the 10 x 40 finder. At low power, which gives best results, the field is particularly striking.

The brightest members of M50 form a heart-shaped figure. A red star mentioned by Admiral Smyth and T. W. Webb is presumably the 8th-magnitude object about 7 minutes of arc south of the cluster's center, but the color was not conspicuous to author Mallas.

On the Harvard Observatory photograph reproduced below are circled four of Messier's open clusters, in Puppis, Canis Major, and Monoceros. Counterclockwise from top right, they are M50, M46, M93, and M41. The very overexposed image is that of Sirius, the brightest star in the sky. Close inspection will reveal many other clusters and nebulae in this rich portion of the southern winter Milky Way.

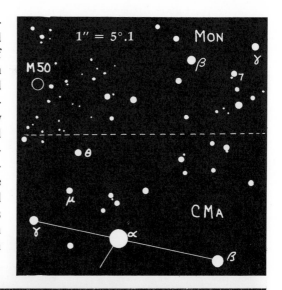

Galaxy in Canes Venatici
M 51
NGC 5194   13h 27m.8   + 47° 27'

Basic data. This is the famous Whirlpool galaxy, with its companion NGC 5195 just 4 minutes of arc to the north. M51 is approximately 11 by 7 minutes in size and has a total magnitude of 8. NGC 5195 is only 3 by 2 minutes and about 1½ magnitudes fainter. M51 is clearly a late-type spiral, while the companion is an irregular system similar to M82. Both lie at a distance from us of roughly 15 million light-years.

The spiral structure of M51 was first seen in 1845 with the 72-inch reflector of the Earl of Rosse. M51 is another example of an Sc galaxy. On photographs taken with large modern reflectors, the entire spiral pattern is dominated by dark dust lanes. The arms can be traced for 1½ turns. One dust lane from M51 crosses in front of NGC 5195 and is visible in the Kreimer photograph.

NGC description. M51: A magnificent object, great spiral nebula. NGC 5195: Bright, pretty small, little extended, very gradually brighter

114

toward the middle, involved in M51.

Visual description. The Whirlpool galaxy is one of the finest objects in the heavens. Author Mallas has seen both M51 and its companion in 7 x 50 binoculars and found them impressive in his 4-inch at low powers. The central part of M51 gave an impression of having texture. Although the drawing contains a weak spiral pattern and a bridge connecting the two galaxies, these fea- 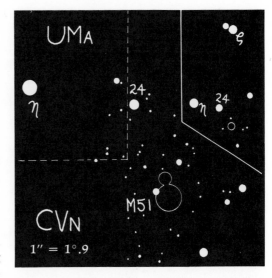 tures may be spurious, and merely a result of the observer's familiarity with photographs.

Under clear desert skies, the spiral pattern and dark rifts could be easily seen with a 12½-inch f/4 reflector. As far as visual detection of spiral arms in M51 is concerned, much depends on an experienced eye and very favorable viewing conditions.

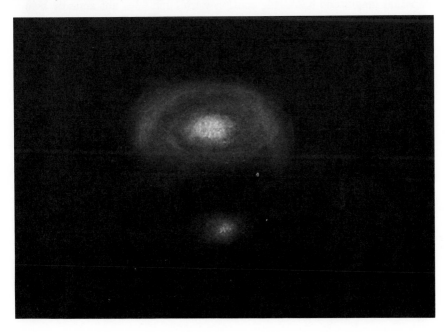

M 52

Galactic cluster in Cassiopeia

NGC 7654   23h 22m.0   + 61° 20′

Basic data. With a total visual magnitude of 7, Messier 52 is about 12 minutes of arc in diameter. It is a rather rich open cluster with roughly 200 members brighter than 15th magnitude. A conspicuous reddish star of visual magnitude 8.2 at the southwestern edge does not actually belong to the cluster, whose members are at least three magnitudes fainter. Recent estimates place M52 about 7,000 light-years away. On September 7, 1774, Messier saw this cluster close to the comet of that year.

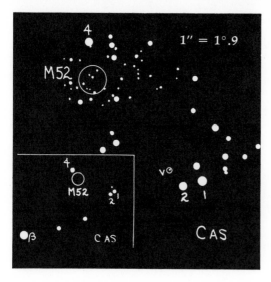

NGC description. Cluster, large, rich, much compressed in the middle, round, stars from 9th to 13th magnitude.

Visual appearance. A beautiful sight in a small telescope. The Mallas 4-inch reveals a great many stars in a distinct pattern: a needle-shaped inner region inside a half circle. M52 is south of 5th-magnitude 4 Cassiopeiae.

M 53

Globular cluster in Coma Berenices

NGC 5024   13h 10m.5   +18° 26′

Basic data. At a distance of about 60,000 light-years, M53's angular diameter of 3.3 minutes of arc corresponds to a linear extent of almost 60 light-years. Visually, its total light is about magnitude 7.6. This globular is approaching at 112 kilometers a second. Like M3, its brighter neighbor in the spring sky, M53's stars are not very concentrated toward the center, as compared with other globulars.

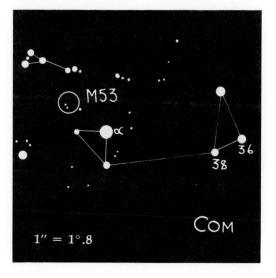

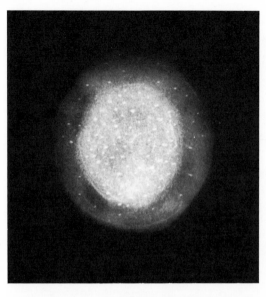

NGC description. Remarkable globular cluster, bright, very compressed, irregularly round. Very much brighter toward the middle; contains 12th-magnitude stars.

Visual appearance. Here is a superb object. It is slightly oval in shape, with a large, bright center. At a magnification of 120, the 4-inch revealed many stars. The central part looked grainy, as if on the threshold of resolution, but the cluster's outer fringes eluded the 4-inch. As shown in the drawing, the bright central region is relatively larger than in M3. The dark areas in the drawing are probably illusions. This globular is rather easy to find, almost exactly a degree northeast of 5th-magnitude Alpha Coma Berenices.

M 54

Globular cluster in Sagittarius

NGC 6715   18h 52m.0   − 30° 32′

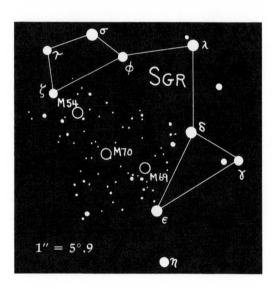

Basic data. The combined light from all the stars of M54 produces a glow roughly equivalent to an 8th-magnitude star. The brightest members are about 14th magnitude and very concentrated toward the cluster's center. The angular size of M54 is variously given as 2 to 5½ minutes of arc. This remote stellar swarm lies at a distance of 50,000 light-years and is receding at 107 kilometers per second.

118

NGC description. Globular cluster, very bright, large, and round. Its brightness increases inward gradually, then suddenly toward the middle. Well resolved, with 15th-magnitude stars and fainter.

Visual appearance. This globular is a splendid sight in the 4-inch. Although it is small in some respects, it gives the visual impression of a compact galactic cluster rather than a globular. The 4-inch refractor reveals some stars in the central region, which is overexposed in the photograph.

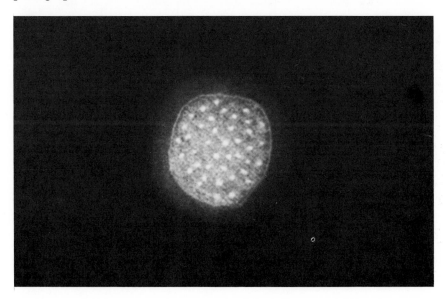

M 55

Globular cluster in Sagittarius

NGC 6809   19h 36m.9   − 31° 03′

Basic data. This 6th-magnitude globular is one of the three objects in the Messier catalogue that were originally discovered by Abbe de Lacaille during his expedition to the Cape of Good Hope in 1751-52.

M55 has an apparent diameter of 15 minutes of arc, which at the cluster's distance of 20,000 light-years corresponds to a linear diameter of roughly 90 light-years. Its stars are quite uniformly distributed, as the photograph shows. M55 is poor in variable stars, only half a dozen being known.

NGC description. Globular, pretty bright, large, round, very rich, very gradually brighter in the middle, stars from 12th to 15th magnitude.

Visual appearance. In the Mallas 4-inch refractor, M55 appears as a large and bright nebulous patch. Its center is more intense than the outer parts and contains a conspicuous bright star. The 4-inch did not resolve the cluster nor reveal a grainy texture.

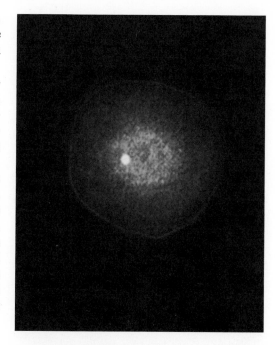

Globular cluster in Lyra

M 56   NGC 6779   19h 14m.6   +30° 05′

Basic data. Located between Gamma Lyrae and Beta Cygni (Albireo), M56 is a compact 8th-magnitude globular star cluster 5 minutes of arc in diameter, about 25 minutes southeast of a 6th-magnitude star. Its distance is about 40,000 light-years, and it is approaching the sun at a speed of 154 kilometers per second. This cluster was first seen by Messier on January 23, 1779. In his small instrument, it must have looked much like a faint comet.

NGC description. Globular cluster, bright, large, irregularly round, gradually very much compressed to-

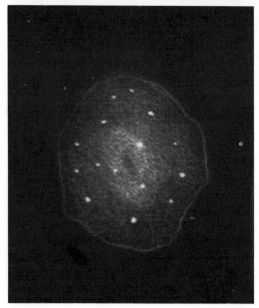

ward the middle, well resolved, stars of 11th to 14th magnitudes.

Visual appearance. An impressive object. In a 4-inch refractor, M56 is a bright, nearly circular glow, in which a few individual stars are seen. Unlike most other globulars, this one has no bright core. It lies in a grand low-power Milky Way field. The Kreimer photograph shows an uneven distribution of stars in the central region, but a larger aperture than a 4-inch would be needed to reveal this effect visually.

M 57

Planetary nebula in Lyra

NGC 6720   18h 51m.7   + 32° 58′

Basic data. The famous Ring nebula was discovered not by Messier but by Antoine Darquier of Toulouse in 1779. It is easily located between Beta and Gamma Lyrae as a patch of light about 1 minute of arc in diameter, about as bright as a 9th-magnitude star.

The nebula is elliptical, its longest and shortest diameters being 74 and 62 seconds of arc, according to visual measurements by Carl Wirtz. On photographs the major axis appears enlarged. The Ring nebula's inner dimensions are about half the outer ones. It is known from the spectrum lines that M57 is slowly expanding. The central star is a difficult object visually, and magnitude estimates disagree. E. E. Barnard found 14.1 and H.

D. Curtis 15.4. Photographically it is about magnitude 14. The distance of the Ring nebula was determined in 1974 by K. M. Cudworth as 4,100 light-years.

NGC description. A magnificent object. Annular nebula, bright, pretty large, considerably extended, in Lyra.

Visual appearance. The Ring nebula is a challenge to the observer. This small object bears magnification well, but the most suitable power depends strongly on sky conditions. The drawing of it on this page was made with a Unitron 4-inch refractor at 250x when conditions were excellent, but normally nothing is gained by using more than 120x.

The drawing shows a star embedded in the brightest part of the nebula and an X where another faint star (not the central one) was suspected. When the seeing is steady, many fine streaks appear. Protracted viewing makes such detail easier.

T. W. Webb long ago noted: "Its light I have often imagined fluctuating and unsteady . . . an illusion arising probably from an aperture too small for the object." In the 4-inch this subjective shimmering occurs for M57 and some other planetary nebulae. This object appears gray with a slightly greenish tinge. A 6-inch shows more green.

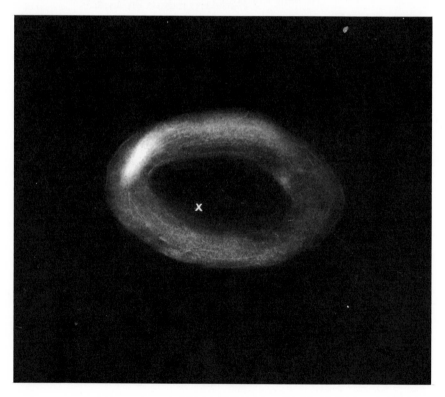

Galaxy in Virgo

M 58   NGC 4579   12h 35m.1   +12° 05′

Basic data. Messier 58 is one of the brightest members of the great Virgo cluster, a swarm of over 1,000 galaxies that spreads across more than 100 square degrees of sky. Although so distant that its light takes about 70 million years to reach us, this is the nearest rich cluster of galaxies.

M58 itself is fairly easily located, since it forms the northern apex of an equilateral triangle with Rho and 20 Virginis. About 4½ by 3½ minutes of arc in size, M58 is equivalent in light to a 9th-magnitude star. The

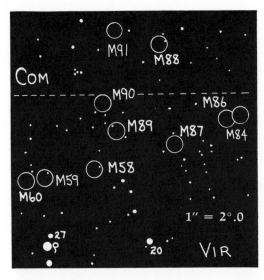

photograph shows its bright central region and faint spiral arms. This system is classed as intermediate between a spiral and a barred spiral.

NGC description. Bright, large, irregularly round, very much brighter in the middle, mottled.

Visual appearance. M58 is visible in the 10 x 40 finder of the 4-inch refractor. The 8th-magnitude star BD +12°2495, about 7 minutes to the west (edge of photo), hampers visual examination. In the 4-inch, the central region of M58 appears lustrous and strongly oval, surrounded by a halo of uneven brightness. Probably the luminous knot toward the northwest in the drawing is identical with the faint star in the spiral arm on the Kreimer photograph. M58 is an interesting object for careful inspection with small apertures.

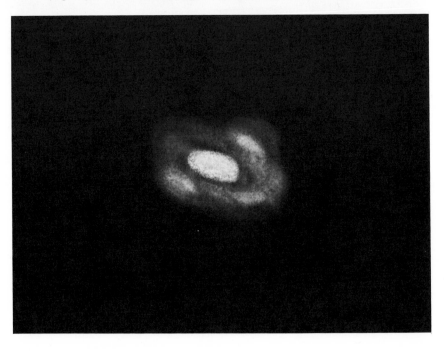

M 59

Galaxy in Virgo

NGC 4621   12h 39m.5   +11° 55′

Basic data. M59 and its neighbor M60 are both elliptical galaxies. They were discovered in April, 1779, by J. G. Köhler of Dresden while observing the comet of that year (Bode-Messier). Messier himself recorded these galaxies a few days later.

In the photograph above, M59 is at the left edge, M60 and fainter NGC 4647 at right, and at top

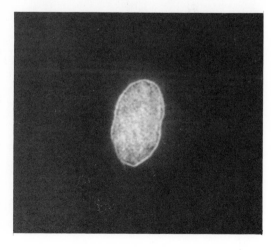

NGC 4638, an elliptical galaxy of about photographic magnitude 12.2.

Though only 2.6 by 1.6 minutes in extent, M59 is as bright as visual magnitude 9½. It belongs to the Virgo cluster and is receding from our own galaxy at about 340 kilometers per second.

NGC description. Bright, pretty large, little extended, very suddenly very much brighter in the middle, two stars preceding [westward].

Visual appearance. This small, hazy, oval patch is difficult to identify because it can be confused with field stars. M59 is a miniature of M32, the brighter of the companions to M31. In the 4-inch refractor, NGC 4638 was seen as a more or less circular gray patch.

Galaxy in Virgo

M 60   NGC 4649   12h 41m.1   +11° 49′

Basic data. This 9th-magnitude elliptical system lies 25 minutes of arc east and a little south of M59. It is about 3 minutes in diameter and slightly elongated. M60 is noteworthy for the very many faint globular clusters surrounding it, visible only on long-exposure photographs with the largest telescopes. The intrinsic luminosity of this galaxy equals nearly 300 million suns. A finder chart for M59 and M60 is under M58.

NGC description. Very bright, pretty large, round, the following [eastern] member of a double nebula.

Visual appearance. The light of M60 fades outward from a bright center to a sharp edge. In the same eyepiece field is NGC 4647, which the catalogues describe as an 11th-magnitude late-type spiral. The 4-inch refractor shows it as a faint fuzzy star.

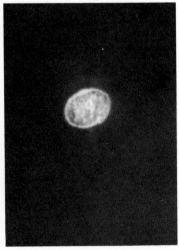

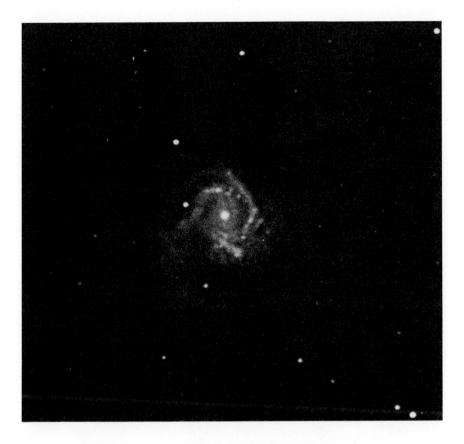

M61

Galaxy in Virgo

NGC 4303   12h 19m.4   + 4° 45'

Basic data. M61 is a spiral galaxy, about 5 minutes of arc in diameter, with low surface brightness. At 10th magnitude, it is much less prominent than its neighbor M49.

The Kreimer photograph reveals clearly the spiral arms and small bright nucleus. We can also recognize the north-south "pseudo bar" through the nucleus, mentioned in the *Hubble Atlas of Galaxies,* which states that this system has some of the characteristics of a barred spiral galaxy.

In his 18¾-inch reflector, Sir John Herschel on one occasion saw M61 as an 11th-magnitude nucleus surrounded by a very faint glow 2 minutes in diameter. Another time he drew it as two overlapping nebulosities, their centers 1½ minutes of arc apart on a northeast-southwest line.

M61, which is probably an outlying member of the great Virgo cluster

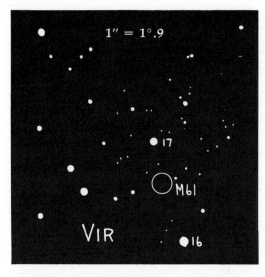

1" = 1°.9

17

M61

VIR

16

of galaxies, has a red shift of about 1,550 kilometers per second.

NGC description. Very bright, very large, very suddenly brighter toward the starlike center, binuclear.

Visual appearance. A fine object for the 4-inch refractor. Inside a large, faint, nearly circular area is a bright elongated center, which does not give the impression of graininess that M81 and some other galaxies do. The starlike nucleus is not seen in the 4-inch. The three brighter patches drawn in the outer region can be identified in the photograph as sections of the spiral arms.

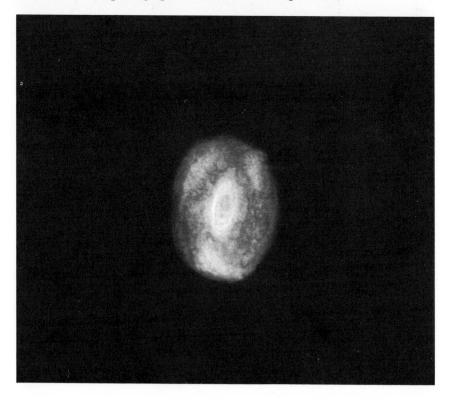

M 62

Globular cluster in Ophiuchus

NGC 6266   16h 58m.1   − 30° 03′

Basic data. With a visual magnitude of 7 and an apparent diameter of 6 minutes of arc, M62 is nearly a twin of M19, which lies 4° to the north. However, M62 is somewhat farther away. In contrast to M19, which is very poor in variable stars, M62 has about two dozen.

NGC description. Remarkable globular, very bright, large, gradually much brighter toward the middle, well resolved, stars

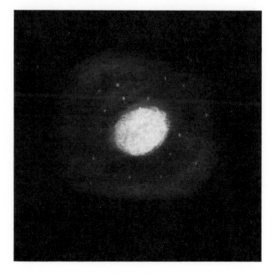

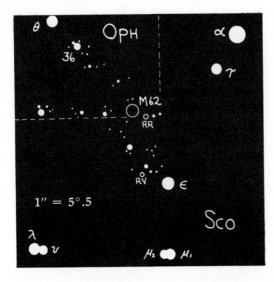

of 14th to 16th magnitude.

Visual appearance. An impressive object. The central region is very bright, compact, and also slightly grainy. Surrounding this is a soft irregular glow with a few foreground stars.

In the photograph, there appear to be fewer stars toward the west, while the burned-out "center" of the cluster is displaced toward the east. This effect, however, was not noticed visually by author Mallas.

Galaxy in Canes Venatici

M 63   NGC 5055   13h 13m.5   + 42° 17′

Basic data. This galaxy was not discovered by Messier, but by his friend Pierre Méchain in 1779. It lies about 6° southwest of M51 (the famous Whirlpool nebula) and 1½° north of the conspicuous row of stars 19, 20, and 23 Canum Venaticorum.

M63 is about 10 by 6 minutes in size and shines with a total light equivalent to an 8½-magnitude star. This spiral galaxy is receding from us at 600 kilometers a second; it is approximately at the same distance as M94.

NGC description. Very bright, large, and pretty much extended in position angle 120°. Very suddenly much brighter in the middle toward a bright nucleus.

Visual appearance. Like several other galaxies in Canes Venatici, M63 is very impressive in small tele-

scopes. As the drawing shows, this system has a strange visual appearance in a 4-inch refractor, with one end more pointed than the other. From the edge inward, the surface brightens slowly, then more rapidly toward the oval, grainy-looking central condensation.

In the Kreimer photograph, the central region is uneven in brightness and texture, agreeing with the visual impression. The spiral arms recognizable in the photograph appear in the drawing as the outlying unresolved soft glow.

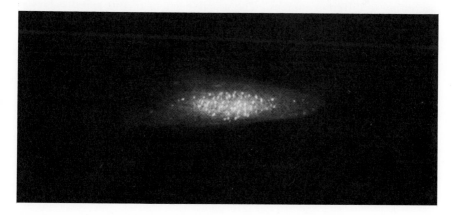

Galaxy in Coma Berenices
M 64   NGC 4826   12h 54m.3   + 21° 57′

Basic data. Messier 64 has a total visual magnitude of 8.8 or 8.1 according to different sources. The angular extent of this spiral galaxy is 6 by 3 minutes on photographs, but it appears smaller visually. It has a red shift of 352 kilometers per second and is perhaps 12 million light-years distant. The *Hubble Atlas of Galaxies* contains two photographs of M64 taken with the 60-inch Mount Wilson reflector and makes particular mention of the soft, smooth, circular spiral arms. Allan Sandage, the atlas' author, classifies this galaxy as an Sb spiral due to the great lane of dust near its center. M64 is often called the Black-eye galaxy because of this dark feature; it is prominent in the Kreimer photograph, which also shows the spiral arms well.

NGC description. Remarkable, very bright and large, greatly elongated in about position angle 120°. Has a brighter middle with a small bright nucleus.

Visual appearance. Because of its details, M64 ranks as one of the finest Messier objects in the 4-inch. There has been disagreement among visual observers as to the detectability of the black eye, which has the reputation of being elusive.

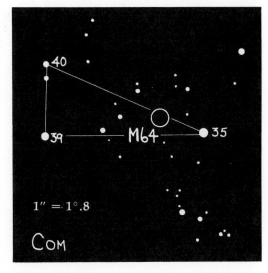

Author Mallas saw the black eye in a 2.4-inch, and found it easy in the 4-inch but subdued in an 8. A 12½-inch showed it at medium but not at low powers.

In the 4-inch, medium power was needed to bring out sharply the features in the drawing, including a seeming cluster of stars northwest of the black eye. This "cluster" is actually a group of bright and dark patches, well seen in the short-exposure picture in the *Hubble Atlas of Galaxies.* Visually, M64 is irregularly shaped, with very uneven brightness and texture.

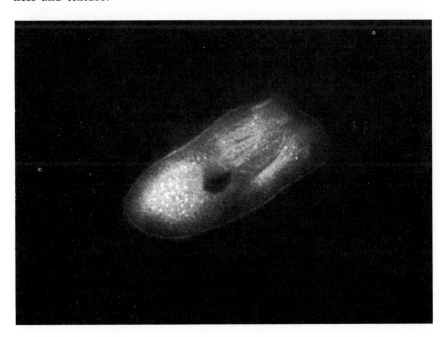

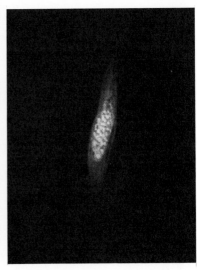

Galaxy in Leo

M 65

NGC 3623   11h 16m.3   +13° 23'

Basic data. Since M65 lies only 20 minutes of arc to the northwest of M66, both galaxies can be seen in the same low-power field. They may be 35 million light-years from us.

In total light, M65 is a match for a star of magnitude 9, and it spans 8 by 1½ minutes. Classed as an Sa or early-type spiral from its photographic rather than its visual appearance, M65 has a prominent central lens and thin, smooth, tightly wound arms.

NGC description. Bright, very large, much extended in position angle 165°, gradually brightening to a bright central nucleus.

Visual appearance. This remarkable object is beautiful in the 4-inch. The oval central region is white and has a very granular appearance, as the drawing shows.

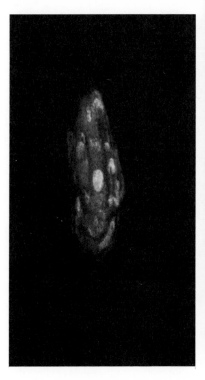

Galaxy in Leo

M 66   NGC 3627   11h 17m.6   +13° 17′

Basic data. M66 slightly outshines its near neighbor M65. Having an angular size of 8 by 2½ minutes, it is as long as M65 but somewhat wider, being seen less nearly edge on. The finder chart is on the facing page, and the star 73 Leonis is of visual magnitude 5.5.

A spiral galaxy of intermediate type, M66 is classified Sb. Photographs with large telescopes show many dark markings due to obscuring dust. Particularly prominent is the great dark lane along the eastern side of this galaxy.

NGC description. Bright, very large; much extended in position angle 150°; much brighter in the middle; two stars northwest.

Visual appearance. This interesting object has a mottled or clumpy appearance, reminiscent of the Orion nebula. Many dark areas could be seen without difficulty in the 4-inch. The central condensation seemed soft and nebulous when well observed. Although the 4-inch shows only the bright inner part of this galaxy, there is a reasonable correspondence between the details in the Mallas drawing and the Kreimer photograph.

M67

Galactic cluster in Cancer

NGC 2682   8h 48m.3   +12° 00′

Basic data. Through binoculars M67 is a 6th-magnitude patch about ¼° across. It is rich, with over 150 members brighter than magnitude 15. M67 is 2,250 light-years away and 11 light-years in diameter. Unlike most open clusters, M67 lies about 1,500 light-years above the plane of our galaxy. It is very old, perhaps 10 billion years.

NGC description. Remarkable cluster, very bright and large, extremely rich, little compressed, stars from 10th to 15th magnitude.

Visual appearance. An easy cluster to resolve. In the 4-inch, the star hues of M67 are predominantly rust, orange, gold, and yellow.

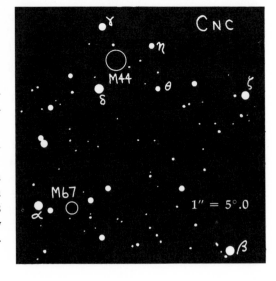

M 68

Globular cluster in Hydra

NGC 4590   12h 36m.8   − 26° 29′

Basic data. Messier 68 is an 8th-magnitude globular cluster discovered in 1780. Photographically it extends 10 minutes of arc, but visually it may seem only half that large. It lies about 40,000 light-years away.

About 40 variable stars have been detected in M68, mostly of the RR Lyrae type. Nearby, but not a member of the cluster, is the Mira-type variable FI Hydrae, period 324 days, which can become as bright as visual magnitude 9. On the photograph, where this star is indicated between

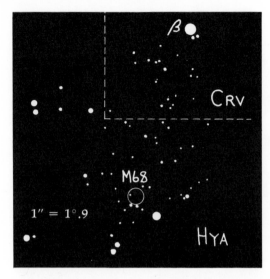

$1'' = 1°.9$

two lines, it is about 100 days past maximum.

NGC description. Globular cluster of stars, large, extremely rich, very compressed, irregularly round. Well resolved, stars of magnitude 12 and fainter.

Visual appearance. A beauty! With medium magnification on the Mallas 4-inch refractor, M68 was oval, with a bright central region surrounded by a fringe that faded outward to a ragged edge. The cluster was unresolved apart from a few of its brightest stars. About ½° southwest is a 5th-magnitude star which looked yellowish in the 4-inch; as the finder chart shows, this is the only bright star near M68.

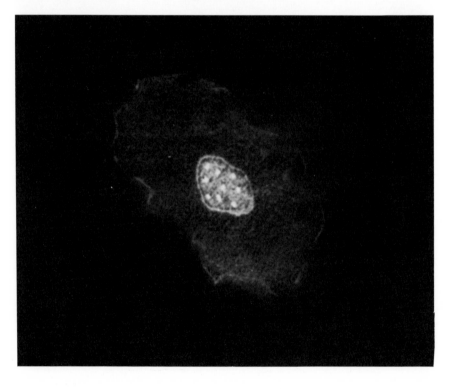

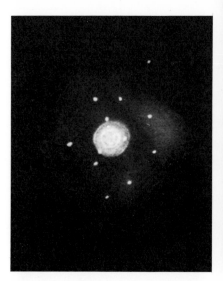

Globular cluster in Sagittarius

M 69

NGC 6637   18h 28m.1   − 32° 23′

Basic data. M69 appears about 9th magnitude visually, has an apparent diameter of 3 minutes of arc, and lies about half as far away as M54. The linear diameter of this cluster is about 70 light-years, and the concentration of stars is about average for a globular.

NGC description. Globular, bright, large, round, well resolved, stars of 14th to 16th magnitude.

Visual appearance. Despite its southerly declination, this is another impressive object, very compact with an intense core. M69 lies close to an 8th-magnitude star that makes low-power viewing difficult. The Kreimer photograph matches the visual impression quite well, showing an outer fringe of stars which are clumped unevenly. These knots may have been mistaken for stars in the 4-inch.

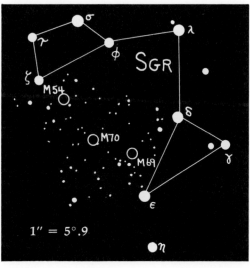

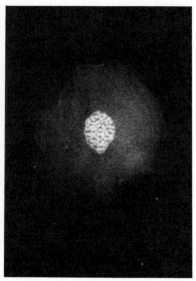

Globular cluster in Sagittarius

M70   NGC 6681   18h 40m.0   — 32° 21′

Basic data. As seen directly, M70 has a diameter of roughly 3 minutes of arc and a visual magnitude of 9. Lying at a distance of about 65,000 light-years, this globular's linear diameter is some 60 light-years. Like its neighbors M54 and M69 in this part of the sky, M70 is receding from us at nearly 200 kilometers a second.

NGC description. Globular, bright, pretty large, round, gradually brightening toward the middle, stars from 14th to 17th magnitude.

Visual appearance. A tight cluster, round but with an irregular outline. The central region is very bright and grainy in the 4-inch. Here is an easy and impressive target for telescopes of much less aperture than the 4-inch.

142

M71

Globular cluster in Sagitta

NGC 6838   19h 51m.5   +18° 39′

Basic data. The total light of this star cluster is roughly that of an 8th-magnitude star. For many years it was regarded by some astronomers as a rather condensed galactic cluster, but in 1946 was clearly recognized as a globular. The apparent diameter of M71 is 6 minutes of arc. Its distance is perhaps only 8,500 light-years, making it one of the nearest globulars. M71 is approaching us at roughly

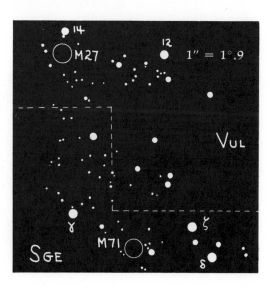

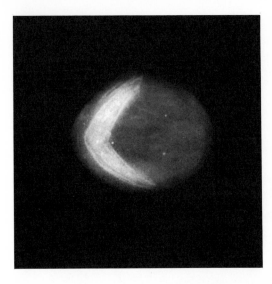

80 kilometers per second.

NGC description. Cluster, very large, very rich, pretty much compressed, stars from 11th to 16th magnitude.

Visual appearance. A beautiful sight even in Mallas' 10 x 40 finder. His 4-inch refractor did not resolve M71, and only a few foreground stars were seen projected against its glow. Visually, M71 is an oval with the brighter side forming a curving "V."

Globular cluster in Aquarius

M72   NGC 6981   20h 50m.7   −12° 44′

Basic data. At 60,000 light-years from us, this 10th-magnitude globular's 2-minute diameter corresponds to 35 light-years. Like M30, it is approaching us at 255 kilometers per second.

NGC description. Globular, pretty bright and large, round, much compressed in the middle, well resolved.

Visual appearance. In the 4-inch, M72 is a very small and nebulous patch of light, with the core the most intense part. The graininess shown in the drawing suggests that this cluster is a loose one, like M4 in Scorpius. The finder chart is opposite.

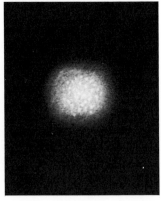

M73

Asterism in Aquarius

NGC 6994   20h 56m.4   −12° 50′

Basic data. Is M73 a true Messier object? Owen Gingerich supplies a translation of Charles Messier's own description of it in 1780: "Cluster of three or four small stars, which resembles a nebula at first glance, containing very little nebulosity; this cluster is located on the parallel [of declination] of the preceding nebulosity [M72]; its position has been determined from the same star [Nu Aquarii]."

Therefore, it seems that M73 is a valid if minor Messier object. Admiral W. H. Smyth saw M73 in 1836 as "a trio of 10th-magnitude stars in a poor field."

NGC description. Cluster, extremely poor, very little compressed, no nebulosity.

Visual appearance. Messier's description matches what was seen in the 4-inch. Moderate magnification shows the quartet, centered in the photograph.

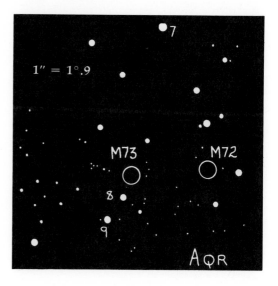

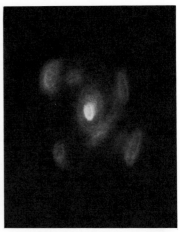

M74   NGC 628   1h 34m.0   +15° 32′

Basic data. M74 is, like M33, classed as an Sc spiral, but it is several times more remote, at 20 million light-years. Therefore it looks smaller (8 minutes in diameter) and fainter (visual magnitude 8.8). A drawing published by Lord Rosse in 1861 shows M74 as a hazy patch with a nucleus strongly resembling an unresolved globular. Hence the object was wrongly called a star cluster in Dreyer's *New General Catalogue*.

NGC description. Globular cluster, faint, very large, round, pretty suddenly much brighter toward the middle, some stars seen.

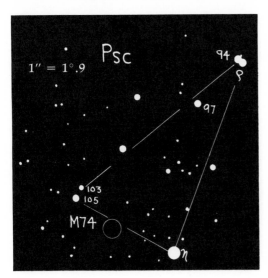

Visual appearance. This is a difficult galaxy for the 4-inch refractor, but it is easily seen in the 10 x 40 finder. A casual observer might miss this object completely, for the central condensation is starlike and the outer parts have very low surface brightness. The most interesting features noted in the 4-inch are the very faint nodules around the center.

146

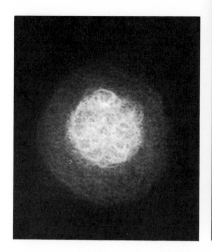

Globular cluster in Sagittarius

M75
NGC 6864   20h 03m.2   −22° 04′

Basic data. Lying near the Capricornus border, M75 is the most remote globular in the Messier catalogue, being 100,000 light-years distant according to Helen Sawyer Hogg. The apparent diameter of M75 is about 5 minutes of arc, and S. van den Bergh has measured its total light as equivalent to a star of visual magnitude 8.7.

About a dozen variable stars have been discovered in this globular, but little is known about them. In the Kreimer photograph, note the extremely compact core and the halo of faint stars.

NGC description. Globular, bright, pretty large, round, very much brighter toward the middle to a much brighter nucleus, partially resolved.

Visual appearance. An intense, nebulous, central region is surrounded by a glow that fades steadily into the sky background. In the 4-inch the appearance of M75 is unusual, with neither stars nor grainy texture.

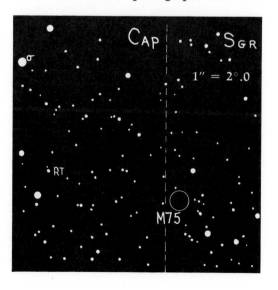

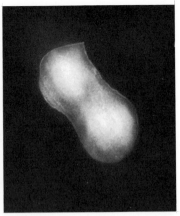

M76

NGC 650-1   1h 38m.8   + 51° 19′

Basic data. M76 appears on photographs as a nebulous bar about 1½ by ¾ minutes of arc in size, extending northeast to southwest, with much fainter patches on either side. Because the ends of the bar are brighter than the middle, William Herschel called M76 a double nebula, which led to the two numbers in the *New General Catalogue*.

Since its total light is equivalent only to a star of 10th or 11th magnitude, M76 is among the faintest objects in the Messier catalogue. The central star of this planetary, in the middle of the bar, is of photographic magnitude 16½. M76 is at a distance of 3,400 light-years.

NGC description. NGC 650 and 651, both very bright, are the preceding and following components of a double nebula.

Visual appearance. A rewarding object in Mallas' 4-inch refractor. M76 is a miniature of the Dumbbell nebula and is more closely described by this name than is M27 in Vulpecula, when both are seen in a small telescope.

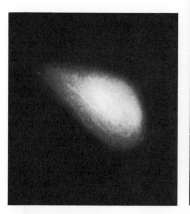

Galaxy in Cetus

M77

NGC 1068   2h 40m.1   −0° 14′

Basic data. Situated far from the Milky Way, M77 is the only Messier object in this part of the sky, but it has many fainter galaxies in its vicinity. According to G. de Vaucouleurs, M77 is 6 minutes of arc in diameter and of total visual magnitude 8.4. M77 is a Seyfert-type galaxy, characterized by a starlike nucleus with an emission-line spectrum, and is perhaps 30 million light-years distant from the earth.

NGC description. Very bright, pretty large, irregularly round, suddenly brighter toward the middle, some stars seen near the nucleus.

Visual appearance. One of the best galaxies for viewing in small apertures. Its irregular shape is beautiful in the 4-inch at low to medium magnifications. The intense central region does not show the granular texture seen in some galaxies. Around it is an uneven glow, which in turn is surrounded by a very feeble haze of light that has not been included in the drawing. There is a good agreement between the visual and photographic characteristics as recorded by the authors of this album.

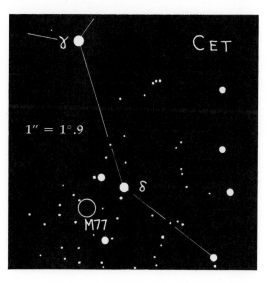

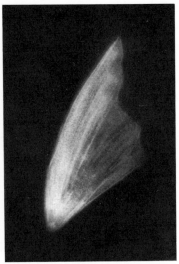

Diffuse nebula in Orion

M78

NGC 2068   5h 44m.2   +0° 02′

Basic data. Not far from Orion's Belt, this 8-by-6-minute patch is part of the great Orion complex of nebulosity. Unlike M42-43, however, which are glowing gas, M78 shines only by reflecting the light of the 10th-magnitude star HD 38563. The photograph shows at lower right (northeast) NGC 2071, close northwest NGC 2067, and to the southwest very faint NGC 2064.

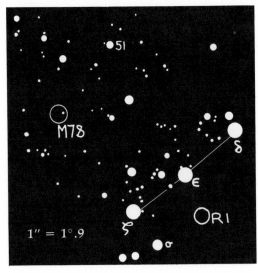

NGC description. A large, bright wisp, gradually much brighter toward a nucleus, three stars involved, mottled.

Visual appearance. In the 4-inch refractor, M78 looks rather like a faint comet with a compact head and short, broad tail, as shown in the drawing. At 60x NGC 2071 is visible as a soft glow. The sky surrounding M78 seems to have a misty sheen, and stars are fewer to the west.

M79

Globular cluster in Lepus

NGC 1904   5h 22m.2   −24° 34′

Basic data. M79 is an 8th-magnitude object about 3 minutes of arc in diameter on a line from Alpha Leporis through Beta and 4° farther. It is ½° northeast of ADS 3954, a 5½-magnitude star with a 7th-magnitude companion at 3 seconds of arc. M79 is about 54,000 light-years from us.

NGC description. Globular cluster, pretty large, extremely rich and compressed, well resolved into stars.

Visual appearance. An impressive globular in the 4-inch, which shows a bright glow with a few stars at its edges. The mottling that Rev. T. W. Webb noted with high powers was not detected by author Mallas. The Kreimer photograph shows M79 to have a compact central region that is surrounded by a straggling haze of stars which range considerably in apparent brightness.

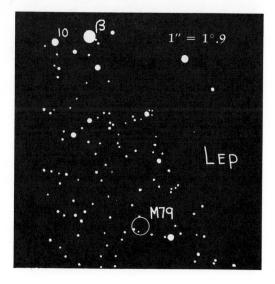

151

Globular Cluster in Scorpius

M 80   NGC 6093   16h 14m.1   − 22° 52′

Basic data. Lying nearly midway between Antares and Beta Scorpii, M80 to users of small telescopes strongly resembles an 8th-magnitude comet about 3 minutes of arc in diameter. It was discovered by P. Méchain on January 4, 1781. Recent estimates of the cluster's distance average about 36,000 light-years.

A 7th-magnitude nova appeared in the center of M80 in May, 1860, and faded to invisibility in a month. This star, T Scorpii, is one of two novae known to have flashed up in a globular cluster (the other occurred in Messier 14 in 1938).

NGC description. Remarkable globular cluster, very bright, large, very much brighter in the middle, readily resolved, contains stars of the 14th magnitude and fainter.

Visual appearance. This splendid object can be detected with only slight optical aid. The 4-inch shows it as round with a bright center. M80 takes magnification well. As may be seen from their photographs, there is a marked

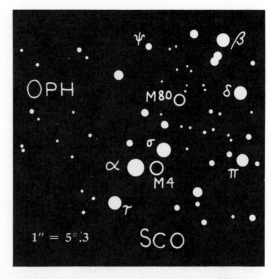

contrast in appearance between M80, with its strong central concentration of stars, and its neighbor M4, which is little concentrated.

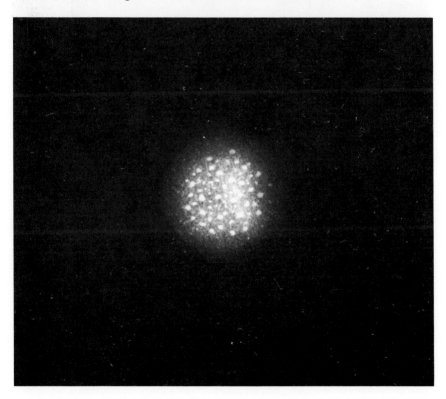

Galaxy in Ursa Major

M 81

NGC 3031   9h 51m.5   + 69° 18'

Basic data. Messier 81 has a total magnitude of about 7.9 visually and 8.4 photographically. Its extreme dimensions on photographs are 21-by-10 minutes of arc. This splendid Sb spiral is about 7,000,000 light-years distant, according to Allan Sandage in 1954. As is to be expected for such a relatively nearby galaxy, the red shift is small: 88 kilometers per second.

NGC description. A remarkable object, extremely bright and extremely large, extended in position angle 156°. It increases in brightness inward, first gradually and then suddenly to a very much brighter center. Bright nucleus.

Visual appearance. In the 4-inch refractor, a beautiful object! As seen in that telescope, M81 has the most strongly granular central region of almost any galaxy. The outer parts are mottled and uneven in brightness and texture. Two other visual characteristics are the fairly sharp outer edge and the bright arcs at the ends of the major axis. Comparison of the drawing, where many fainter details are exaggerated intentionally, with the photograph indicates that these arcs are portions of spiral arms. In the finder chart, R is a

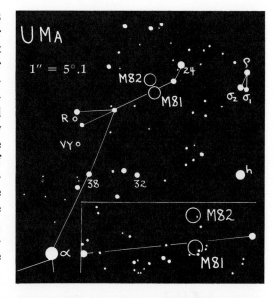

Mira-type variable with a period of 302 days and a range from 7th to 13th magnitude. VY is a 6th-magnitude irregular variable of small range.

M82

Galaxy in Ursa Major

NGC 3034   9h 51m.9   +69° 56′

Basic data. An irregular galaxy about 9-by-4 minutes of arc in extent, M82 is of magnitude 8.8 visually and 9.4 photographically. About the same distance from us as its neighbor M81, it also has a small spectral red shift, 322 kilometers per second. The finder chart is on the previous page.

NGC description. Very bright and very large. An extended ray.

Visual appearance. A gem! In a low-power field, it forms a beautiful pair with M81. In shape and color M82 is a silver sliver, with its brightest part off-center as in the drawing. The dark absorption band seen in the photograph was not detected in the 4-inch. This telescope showed the galaxy to be highly uneven in brightness but with little or none of the grainy texture seen in M81.

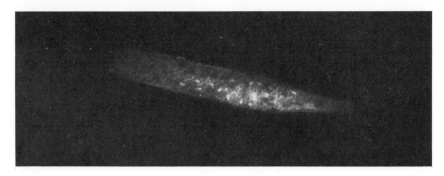

M83

Galaxy in Hydra

NGC 5236   13h 34m.3   −29° 37′

Basic data. M83 was discovered by the French astronomer Lacaille during an expedition to South Africa in 1751-52. This 7th-magnitude galaxy is seen nearly face on. Photographically it appears about 10-by-8 minutes of arc in angular size with a bright nucleus. G. de Vaucouleurs, who gives the distance as about 8 million light-years, classified it as intermediate between ordinary spirals and barred spirals. Four super-novae have been discovered in this system, in 1923, 1950, 1957, and 1968.

NGC description. A very remarkable object. William and John Herschel found it very bright, very large, elongated in position angle 55°,

157

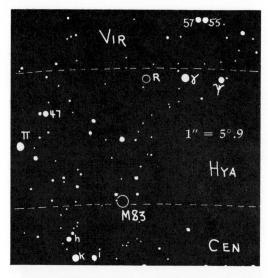

very suddenly much brighter toward a central nucleus. Seen as a three-branched spiral by Leavenworth [with the 26-inch refractor of Leander McCormick Observatory].

Visual appearance. The 4-inch revealed enough to suggest that M83 must be a magnificent object for the Southern Hemisphere observers, who can see it high in the sky. The Mallas drawing shows only the bright elliptical inner region, which seems to have some graininess near its extremities, perhaps due to the presence of the spiral arms revealed in the Kreimer photograph.

Galaxy in Virgo

M84   NGC 4374   12h 22m.6   +13° 10′

Basic data. This elliptical-type galaxy is round, with an apparent diameter of about 2 minutes of arc. It has a total visual magnitude of 9.3 and is conspicuous at the left edge in the 12½-inch reflector photograph above. M84 is receding from us at nearly 900 kilometers per second.

Eight other galaxies in the Virgo region can be identified in the photograph, the brightest being M86, below and left of center. In the following listing, visual magnitudes and dimensions in minutes of arc are given in parentheses: at upper left is the edgewise spiral NGC 4388 (11.3, 5 x 1); below it (northward) and nearly starlike is NGC 4387 (11.8, 1 x ½); at top center is the barred spiral NGC 4413 (11.8, 2 x 1); to its lower right is NGC 4425 (11.6, 2 x ½); below M86 is dim edgewise 4402 (11.5, 4 x 1); brightest at lower right is NGC 4438 (9.8, 8 x 3); 4 minutes of arc to the north is the elliptical NGC 4435 (10.5, 1 x ¾). This 12-minute exposure was taken March 21, 1966.

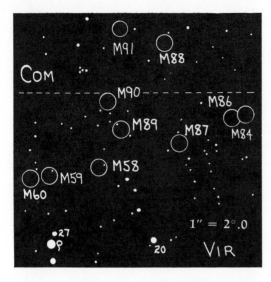

NGC description. Very bright, pretty large, round, pretty suddenly brighter in the middle, mottled.

Visual appearance. This beautiful sight is an easy target for the 4-inch and is even visible in the refractor's 10x 40-mm. finder. More conspicuous visually than M86, which is seen in the same field, M84 has a bright center and fades smoothly to a diffuse edge. It looks like an unresolved globular cluster. In the drawing of M84 at left below, the dark areas on the disk of the galaxy are probably illusions caused by eye strain.

The drawing at right below is of M85, which has nearly the same right ascension as M84, but lies 5°.3 to the north, at the edge of the Coma-Virgo region of galaxies.

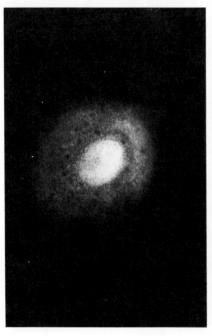
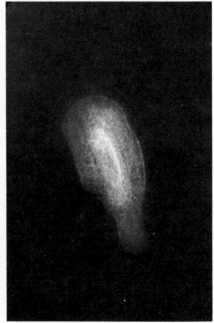

M 85

Galaxy in Coma Berenices

NGC 4382   12h 22m.8   +18° 28'

Basic data. This elliptical galaxy is about 4-by-2 minutes of arc in size and has a total brightness equivalent to a star of visual magnitude 8.9. Only 7 minutes to the east is NGC 4394, a barred spiral that is smaller and fainter than M85. Both are receding from us at approximately 700 kilometers per second.

NGC description. Very bright, pretty large, round, with a bright middle, star north preceding [west].

Visual appearance. In the 4-inch refractor of author Mallas, this beautiful galaxy is a somewhat irregular oval with a smooth texture. The star mentioned in the NGC is shown in the photograph but omitted from the drawing (on the facing page); actually it lies northeast, not northwest, of M85. Perhaps its presence distorted the visual outline. NGC 4394 is visually a silvery oval.

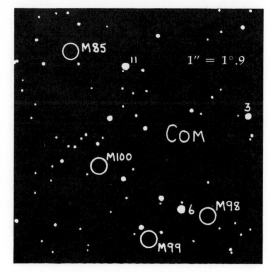

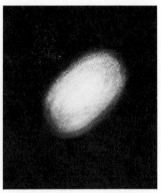

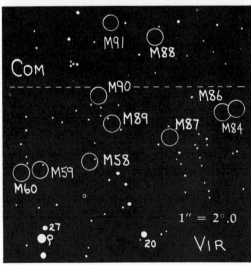

Basic data. M86 is a giant elliptical with a total light equivalent to a 9.2-magnitude star. Oval in form, this galaxy is about 2-by-1 minute of arc. Note the very faint companion system 1½ minutes northeast of the center of M86.

NGC description. Very bright, large, round, gradually brighter in the middle to a nucleus, mottled.

Visual appearance. In a small telescope this galaxy is impressive. Its brightness drops off from the center to a rather sharp edge (contrast the drawing of M86 with that of M84). North of M86 in the photograph is NGC 4402, an 11½-magnitude spiral with a pronounced dust lane, but this object was not seen in the 4-inch.

M 87

Galaxy in Virgo

NGC 4486   12h 28m.3   +12° 40′

Basic data. This is one of the most unusual and important galaxies in the sky. It is the brightest member of the Virgo cluster and is also one of the intrinsically brightest galaxies known, with an absolute photographic magnitude of —20.3. Visually it is of apparent magnitude 8.6 and has a diameter of about 2 minutes of arc. To the southwest, at upper left in the photograph, is an 11th-magnitude elliptical, NGC 4478, about 1.1 minutes of arc across.

M87 is peculiar in being surrounded by a very large number of faint globular star clusters. Over 500 are shown on long-exposure photographs with the biggest telescopes. Also, there is a mysterious bright jet extending northwestward from the nucleus of the galaxy for about 5,000 light-years. The existence of this jet has been known for half a century, and it is shown in the photograph by Kreimer.

In 1954, W. Baade and R. Minkowski identified M87 with a source of

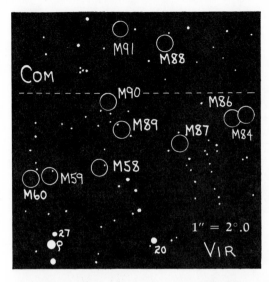

very strong radio emission known as Virgo A. M87 is also a source of X-rays.

NGC description. Very bright and large, round, much brighter in the middle, 3rd of three [easternmost].

Visual appearance. M87 has the typical characteristics of an elliptical galaxy. Its image is smooth and featureless, and the brightness diminishes uniformly from the center into the sky background. The late Otto Struve told of having seen the jet with the 100-inch reflector at Mount Wilson Observatory in California.

M88

Galaxy in Coma Berenices

NGC 4501   12h 29m.5   +14° 42'

Basic data. Messier 88 is one of the visually more rewarding galaxies in the Coma-Virgo region. It resembles the Andromeda nebula as depicted on a very small-scale photograph.

This spiral has an elongated shape, about 5 minutes of arc long and 2½ wide. Its total light equals a 9th-magnitude star. M88 lies at a distance of about 40 million light-years. The finder chart is opposite.

NGC description. Bright, very large, very much extended.

Visual appearance. This object is grand in the 4-inch Mallas refractor. The surface texture is smooth, but the brightness is very uneven, as shown in the drawing. The core appears stellar and is surrounded by a soft glow. The outer parts of M88 are not visible in a small telescope, but the photograph reveals extensive spiral structure.

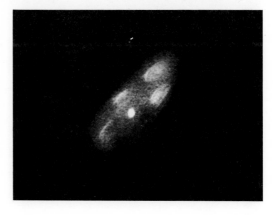

M 89

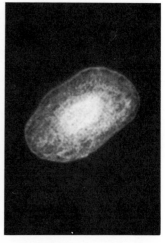

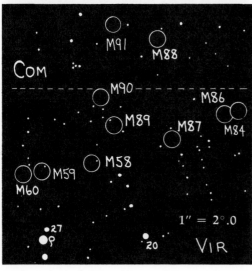

Basic data. This elliptical galaxy (near bottom of photograph) appears as a round patch about 2 minutes of arc in diameter and glows at magnitude 9½. Messier discovered M89 on March 18, 1781.

NGC description. Pretty bright and small, round, gradually much brighter toward the middle.

Visual appearance. In the 4-inch, M89 seems slightly oval, somewhat resembling M87. The brightness increases toward the central core, which does not appear stellar. Visually, the texture is smooth, and the edges blend into the sky background. The 10 x 40 finder shows M89 unmistakably.

Galaxy in Virgo

M 90   NGC 4569   12h 34m.3   +13° 26′

Basic data. Here is a large spiral that spans 7-by-3 minutes of arc. The photograph shows the conspicuous dust lanes extending toward the bright core of this 9th-magnitude galaxy. The total mass of M90 has been estimated as 79 billion suns, from a study of the galaxy's rotation.

NGC description. Pretty large, brighter in the middle toward a nucleus.

Visual appearance. As sketched, M90 is typical of many galaxies seen on a steady dark night. The bright central region does not appear sharp; also the texture is not as smooth as M89 nor do the edges blend gradually into the sky. The outer regions in the photograph were not detected visually.

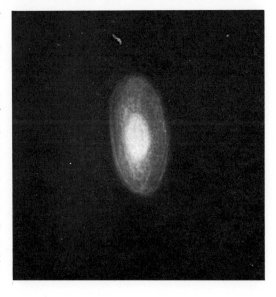

Galaxy in Coma Berenices
M 91
NGC 4548   12h 32m.9   +14° 46′

Basic data. For many years M91 was a "missing" Messier object. It was even believed that the observation of 1781 referred to an unrecognized comet.

Owen Gingerich has suggested that M91 duplicates M58, but W. C. Williams, Fort Worth, Texas, suggests that M91 may be NGC 4548, a 9½-magnitude barred spiral seen nearly face on and about 4 minutes of

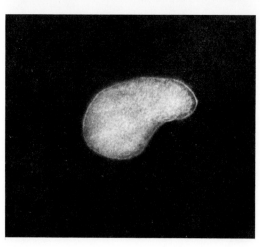

arc in diameter. See the finder chart under M89.

NGC description. Bright, large, little extended, slightly brighter in the middle.

Visual appearance. This galaxy is beautiful even with low power. At medium magnification, its irregularly oval outline and sharp curving extension can be discerned. This extension may be part of the galaxy's bar.

Globular Cluster in Hercules

M 92

NGC 6341   17h 15m.6   +43° 12′

Basic data. Though very nearly as bright as M13 in the same constellation, the globular cluster M92 may look only half as large. M92 contains only about a dozen RR Lyrae variables, the stars that are the primary indicator of a globular's distance. From observations of these, this cluster has been found to lie about 28,000 light-years from us and to be about 100 light-years across.

1" = 2°.0

HER

NGC description. Globular cluster of stars, very bright, very large, extremely compressed in the middle, well resolved, small [faint] stars.

Visual appearance. A grand object, visible with the slightest optical aid. In the 4-inch, M92 is partially resolved at 214x. The visual impression is most unusual, with many stars seen in the bright central region. Surrounding this is a fainter glow that is also star-studded.

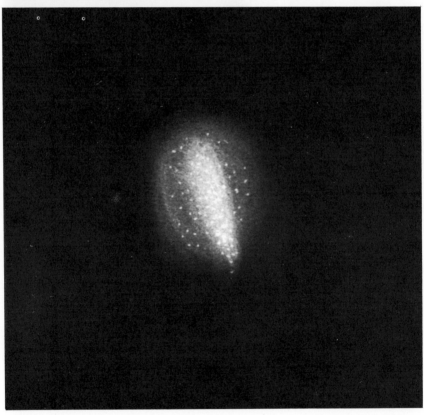

M 93

Galactic Cluster in Puppis

NGC 2447   7h 42m.4   − 23° 45′

Basic data. This rich cluster contains 60 members brighter than magnitude 13½. They cover an area about 18 minutes of arc in diameter and are concentrated toward the center (in contradiction to the NGC characterization). The total magnitude of M93 is about 6.

NGC description. Cluster, large, pretty rich, little compressed, with 8th- to 13th-magnitude stars.

Visual appearance. A glorious view. Author Mallas sees the cluster as triangular in shape, with many colored stars. In the 4-inch at low and medium power, M93 is a compact swarm of stellar jewels. The surroundings of this cluster are grand for sweeping at low power. For the cluster itself, use low to medium power.

171

M 94

Galaxy in Canes Venatici

NGC 4736   12h 48m.6   +41° 23'

Basic data. This bright galaxy was discovered by Pierre Méchain in the year 1781. It is situated 3° north and slightly west of Cor Caroli. Messier 94 spans about 7-by-3 minutes of arc and has a total visual magnitude of 7½. Its distance from us is about 14,500,000 light-years.

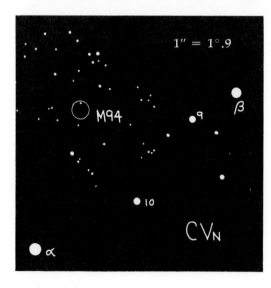

The Kreimer photograph reveals the complex inner details of this spiral galaxy. This structure is overexposed on long-duration photographs with large telescopes, as the 200-inch plate opposite shows.

At the very center of M94 in the Kreimer picture is a bright circular disk, about half a minute of arc in diameter and devoid of spiral structure. Surrounding this hub is a slightly fainter ring made up of tightly wound spiral arms.

Outside this ring is a much larger area filled with many faint fragments of arms. Only a hint of them appears in Kreimer's photograph, but they are conspicuous in the 200-inch picture, from the *Hubble Atlas of Galaxies,* 1961.

The inset reveals a fourth major feature of M94 — a very faint surrounding ring some 15 minutes across. It has a sharp inner edge, but fades into the sky.

NGC description. Very bright, large, irregularly round; very suddenly much brighter toward the middle to a bright nucleus; mottled.

Visual appearance. A grand object! There is a rapid brightening toward a brilliant center, which does not look like a star. The sketch shows an extension seen with the 4-inch refractor; it is probably a segment of one of the brighter spiral arms.

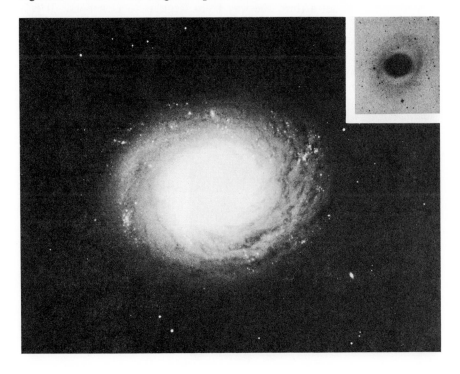

173

M 95

Galaxy in Leo

NGC 3351   10h 41m.3   +11° 58'

Basic data. M95 is one of a trio of 9th-magnitude Messier galaxies located in Leo about a third the way from Regulus to Denebola. It is a barred spiral with a nearly circular arm coming off each end of the bar and winding around almost to the other end. The diameter of the ring is 3 minutes of arc, the bar 1.4-by-0.3 minutes. This galaxy lies at a distance of about 25 million light-years and is receding from us at approximately 640 kilometers a second. The finder chart is opposite.

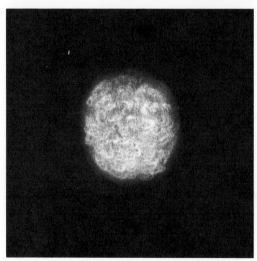

NGC description. Bright, large, round, gradually much brighter toward a nucleus.

Visual appearance. In the 4-inch refractor of author Mallas, M95 appeared as a circular gray patch of uneven brightness. While the bar stands out quite prominently in the photograph, it was not seen in the 4-inch. Only the galaxy's central condensation was visible, as the drawing shows.

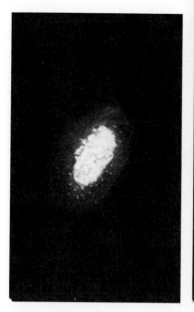

Galaxy in Leo

M 96

NGC 3368   10h 44m.2   +12° 05′

Basic data. This galaxy is a larger spiral than M95, which lies less than a degree to the west. Measuring 6-by-4 minutes of arc, M96 is about the same distance from us as its neighbors.

NGC description. Very bright and large, little extended, very suddenly much brighter in the middle, mottled.

Visual appearance. Its color is silver-gray, and the galaxy has an intense central region. Some grainy texture is seen throughout the oval image. The photograph, however, shows M96 to be smooth with bright and dark lanes. The visual impression refers only to the central region and not the spiral arms.

M 97

Planetary nebula in Ursa Major

NGC 3587   11h 12m.0   + 55° 18′

Basic data. M97 has a total visual magnitude of about 11, though it is harder to see than this might suggest, because its light is spread over an area about 3 minutes of arc across. This planetary is about 1½ light-years in diameter, and lies at a distance of 2,600 light-years, according to K. M. Cudworth.

M97 became known as the Owl nebula from its appearance on a drawing made with Lord Rosse's 72-inch telescope in 1848. In the photograph note the 14th-magnitude central star; above and right of it (to the southeast) is a similarly bright star with a tiny wisp of nebulosity.

NGC description. Very remarkable planetary nebula, very bright and large, round, 150 seconds in diameter. It brightens toward the middle very gradually, then suddenly.

Visual appearance. At 120x in the 4-inch refractor of author John Mallas, M97 appears as a rather large gray oval. It is practically featureless, though there is a slight indication of two dark areas (the Owl's eyes), which are exaggerated in the drawing.

At Covina, California, Mallas was never certain if these areas were seen; in the desert skies of Arizona they were definitely observed, though with difficulty. (Curiously, the eyes are absent from John Herschel's drawing with his 18¾-inch reflector, published in 1833.)

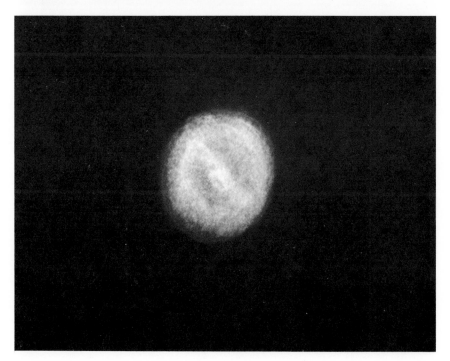

M 98

Galaxy in Coma Berenices

NGC 4192   12h 11m.3   +15° 11′

Basic data. M98 is about 8-by-2 minutes of arc in apparent size and shines at magnitude 9.7. This is an Sb-type spiral with a small, bright nuclear region and well-developed arms. It lies approximately 37 million light-years away. In the photograph by author Kreimer, note the many patches of obscuring dust. M98 forms a pair with the small and faint galaxy NGC 4186. This spiral lies 12 minutes to the southeast.

NGC description. Bright, very large, extended along position angle 152°, suddenly very much brighter toward the middle.

Visual appearance. In the 4-inch refractor, M98 is grainy and mottled like a globular cluster, but with some bright knots superimposed. The drawing shows only the brightest portions of what the photograph reveals to be spiral arms; the 4-inch did not reveal the small bright core. The 5th-magnitude star 6 Comae Berenices lies only 0°.5 to the east.

Galaxy in Coma Berenices

M 99   NGC 4254   12h 16m.3   +14° 42'

Basic data. This nearly face-on spiral appears as a round glow about 4 minutes of arc across. Its total visual magnitude is about the same as that of M98. The spiral arms of M99 are perhaps 5,000 light-years wide, and toward the center of this galaxy are many dust lanes. M99 is receding at 2,400 kilometers a second. The finder chart is opposite; the Kreimer photograph is on the next page.

NGC description. Very remarkable. William and John Herschel called it bright, large, round, gradually brighter in the middle, mottled. F. P. Leavenworth and Lord Rosse saw it as a three-branched spiral.

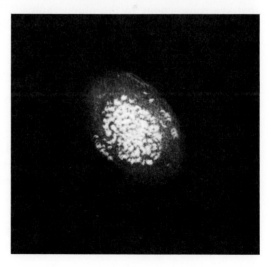

Visual appearance. An impressive sight. The drawing portrays only the large central region, where there are many bright knots. Away from the center these knots become fewer and suggest parts of the spiral arms that are conspicuous in the photograph. While examining M99 with his 4-inch, Mallas had the impression that a small increase in aperture would show much more detail.

M100

Galaxy in Coma Berenices

NGC 4321   12h 20m.4   +16° 06′

Basic data. Like M99, this is also an Sc-type spiral seen nearly face on. About 5 minutes of arc in diameter, M100 has a total magnitude of 9.2, making it one of the brightest spirals in the Virgo-Coma Berenices swarm. The photograph shows two prominent arms and several fainter ones. The distance of this galaxy has been estimated as 40 million light-years.

NGC description. Very remarkable. Pretty faint, very large, round, very gradually then suddenly brighter toward the middle and to a mottled

nucleus. With the 26-inch Leander McCormick refractor, Leavenworth saw M100 as a two-branched spiral.

Visual appearance. A very diffuse, curdled patch of light surrounds a star-like core. To Mallas it seemed like a miniature M33. M100 was usually difficult with the 4-inch refractor. The visual impression matches the photograph quite well. The finder chart is with M98.

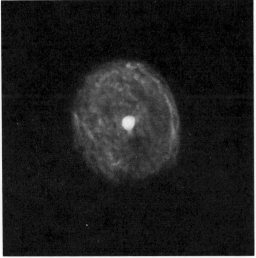

M101 (M102)

Basic data. Through an 18th-century error, this galaxy was also called M102. It is a very large, bright spiral seen face on. Appearing 22 minutes in diameter on photographs, it has a total light equivalent to an 8th-magnitude star. Its distance is about 15 million light-years.

M101 is a late-type spiral of class Sc, with narrow spiral arms only about 900 light-years across and containing many hot, blue stars.

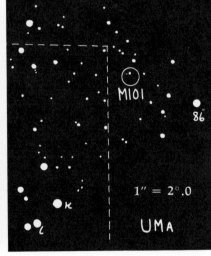

NGC description. Pretty bright, very large, irregularly round. Gradually, then suddenly, much brighter toward a small, bright central nucleus.

Visual appearance. A beautiful object. In the Mallas 4-inch it appears only about half as large as on photographs. Low powers are best. The clearly seen central region has a fluffy texture and a silvery hue. Surrounding this area is a soft sheen containing some nebulous patches.

M 103
Galactic Cluster in Cassiopeia
NGC 581   1h 29m.9   + 60° 27′

Basic data. The apparent diameter of M103 is about 6 minutes of arc, which at the cluster's distance of 8,000 light-years is equivalent to a linear diameter of about 14 light-years. Though this open cluster has a total magnitude of 6.2, its stars are quite scattered. M103 contains about 40 members brighter than magnitude 14. Dominating the cluster, but not a member of it, is the double star Σ131, which has components of magnitude 7.3 and 9.9 separated by 13.8 seconds of arc.

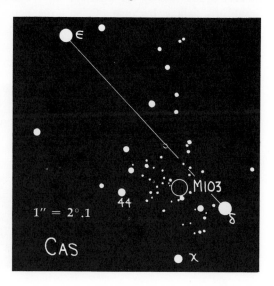

NGC description. Cluster, pretty large, bright, round, rich, stars of 10th and 11th magnitude.

Visual appearance. A grand view! The stars form an arrowhead which is also seen in the photograph. A 10 x 40 finder resolved the cluster, but the 4-inch showed the fainter stars, many of them colored.

M104

Galaxy in Virgo

NGC 4594   12h 37m.3   −11° 21′

Basic data. As bright as magnitude 8 and as large as 6-by-2 minutes of arc, M104 is easy for amateur telescopes. It is widely known as the Sombrero galaxy because of its large central bulge and dark rim of obscuring dust, both well seen in the photograph. This spiral galaxy is nearly edge on, its equatorial plane being tipped only 6° to our line of sight.

NGC description. Remarkable, very bright and large, extremely extended in position angle 92°, very suddenly much brighter toward a central nucleus.

Visual appearance. This is a beautiful object in a 4-inch for the well-trained eye. Until the observer has gained experience, however, the Sombrero galaxy may look as featureless as the great Andromeda galaxy. The central bulge seemed smooth-textured to author Mallas, who saw a faintly luminous border around it and the east-west extensions. Although the "hat brim" is difficult to study visually, it appears to have a curdled texture, which is exaggerated in the drawing. The dark dust lane was not seen.

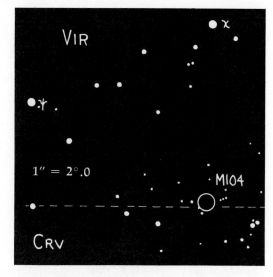

Galaxy in Leo

M105

NGC 3379   10h 45m.2   +12° 51′

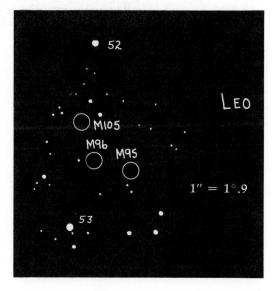

Basic data. M105 is an elliptical system and another member of the Leo group. Only 2 minutes of arc in diameter, it has one of the smallest apparent diameters of the objects in the Messier catalogue.

NGC description. Very bright, considerably large, round, suddenly brighter in the middle, mottled.

Visual appearance. This is the brightest of three galaxies in a low-power field. M105 has a very soft, nebulous texture, like

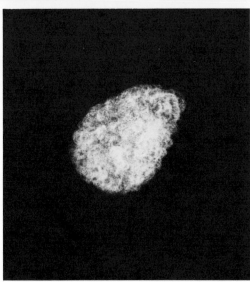

M32 or M87, with a slow brightening toward the center. In good seeing, the image looks like an unresolved globular cluster — an unusual appearance for an elliptical galaxy.

M105 is at left in the photograph. Below center is NGC 3384, a gray circular patch in the 4-inch refractor. The third object, NGC 3389, is the most difficult. During moments of steady seeing it appears as a small, soft, oval film of light.

M 106

Galaxy in Canes Venatici

NGC 4258   12h 16m.5   +47° 35′

Basic data. The original lists by Charles Messier included 103 entries, but several other galaxies and one globular cluster were also logged by him and his colleague Pierre Méchain. Just south of the Big Dipper are two of these galaxies, M106 added by Helen Sawyer Hogg in 1947, and M109 added by Owen Gingerich in 1953.

M106 is a spiral whose faint outlying parts cover an area of 19-by-8 minutes of arc on long-exposure photographs. Its total light is equivalent to an 8th-magnitude star. This galaxy appears strongly foreshortened, since its central plane is tilted only 25° to our line of sight.

This spiral has been known since the 1950's as a radio emission source with an estimated extent of 31-by-18 minutes, considerably larger than the optical size. M106, at a distance of some 25 million light-years, is receding from us at ap-

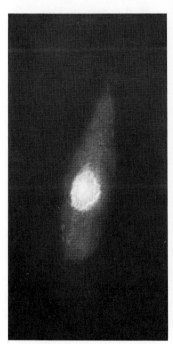

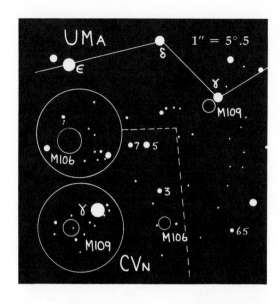

proximately 300 miles per second. Its mass is roughly 100 billion times the sun's.

NGC description. Very bright and large, very much extended north-south, suddenly brighter in the middle to a bright nucleus.

Visual appearance. This grand object is visible in the 10 x 40 finder of the 4-inch refractor. As the drawing shows, there is a fat central knot of fuzzy light. Medium magnifications are helpful for seeing details.

M 107   Globular Cluster in Ophiuchus
NGC 6171   16h 29m.7   −12° 57′

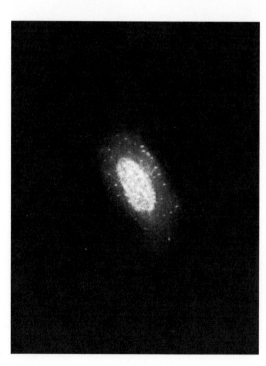

Basic data. This 9th-magnitude globular cluster was discovered by Messier's friend P. Méchain in April, 1782. It was Helen Hogg's suggestion that this object be included in the catalogue along with M105 and M106.

Visually M107 seems about 3 minutes of arc in diameter, though long-exposure photographs show it to be three times that size. At a distance of about 10 thousand light-years, it is some 50 light-years in diameter. It is approaching us at 147 kilometers per second.

NGC description. Glob-

ular, large, very much compressed, round, well resolved.

Visual appearance. Although M107 is rather dim, it is an easy target. It is impressive in the 4-inch at medium magnification. Some individual stars are seen around a flattened unresolved core that is on the threshold of graininess. The fainter halo region appears granular, suggesting that a larger aperture would resolve M107 into many more stars.

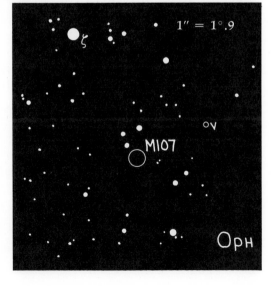

$1'' = 1°.9$

ζ

ο ν

M107

Oph

M 108

Galaxy in Ursa Major

NGC 3556   11h 08m.7   +55° 57'

Basic data. Seen nearly edgewise, 10th-magnitude M108 is very elongated, 8-by-1 minutes of arc. With greatly foreshortened spiral arms and no pronounced nuclear bulge, it is an Sc-type spiral, about 25 million light-years away and receding from us at some 760 kilometers a second.

NGC description. Quite bright, very large, very much extended at position angle 79°, becoming brighter in the middle, mottled.

Visual appearance. A silver-white beauty for small telescopes, saucer-shaped and fairly well defined. The central region is quite bright and irregular, surrounded by light and dark nodules. Mallas found it difficult to match the drawing and the photograph, so it is not certain how much of this galaxy was seen in the 4-inch. The finder chart is with M97.

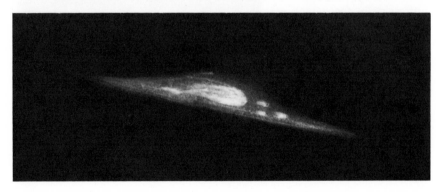

Galaxy in Ursa Major

M 109   NGC 3992   11h 55m.0   + 53° 39′

Basic data. This conspicuous barred spiral lies ⅔° southeast of Gamma Ursae Majoris. It is about 7-by-4 minutes of arc on long-exposure photographs. Through its bright central core extends a stubby bar from the ends of which trail narrow, sharp spiral arms. The total magnitude of M109 is about 9½.

A supernova that appeared in this barred spiral in 1956 briefly attained photographic magnitude 11.2. At the distance of Alpha Centauri, the supernova would have been as brilliant as the full moon!

NGC description. Quite bright, very large, pretty much extended, suddenly brighter in the middle to a bright mottled nucleus.

Visual appearance. A splendid galaxy for small apertures, though only the brighter central region can be seen. It is pear-shaped, with a strong suspicion of a granular texture, and close to a faint star whose glow obliterates the outer regions. The finder chart is with M106.

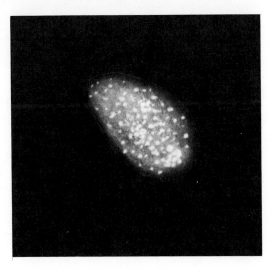

191

Galaxy in Andromeda

M110   NGC 205   0h 37m.6   +41° 25'

Basic data. This most recent addition to the Messier catalogue (and hopefully the last) was proposed in 1967 by K. Glyn Jones. NGC 205 is depicted together with M32 on a Messier drawing of the great Andromeda galaxy, M31, which was published in 1807. A label on the drawing indicates that Messier first observed NGC 205 in 1773, 10 years before Caroline Herschel did, although she has been generally credited with the discovery.

However, Jones found that H. L. D'Arrest in his catalogue of 1867 (a predecessor of the NGC), credits the discovery of NGC 205 to Messier (and later to Caroline Herschel). This is strong support for including this object in modern Messier listings.

Intrinsically NGC 205 is 1/85 as bright as M31. Together with M32, all three of these galaxies lie at a distance from us of approximately 2.2 million light-years.

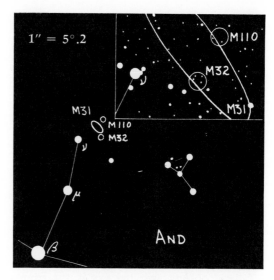

M110 shines at 8th magnitude and is 10 minutes of arc long and about half as wide. The Kreimer photograph, which includes a bit of M31 at upper right, was taken with the 16-inch f/8 Newtonian reflector of Lines Observatory at Mayer, Arizona. Just discernible within the glow of NGC 205 are lanes of dust on the periphery of the intense central region.

NGC description. Very bright and large, much elongated in position angle 165°, very gradually brightening to a very much brighter middle.

Visual appearance. An impressive sight in the 4-inch refractor. It is almost uniform in luster, with a brighter middle. Near the top of the drawing is a curious soft extension that was near the limit of perception in the 4-inch. Other observers, however, do not mention any irregularities in the shape of M110.

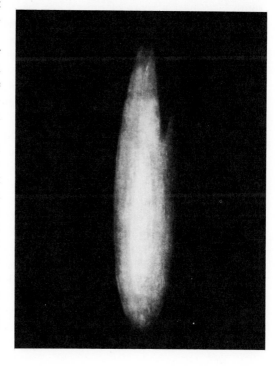

Hints for Beginning Observers

OWEN GINGERICH

Harvard-Smithsonian Center for Astrophysics

O F THE 110 ENTRIES in this Messier catalogue there are: 39 gal-axies, 29 globular clusters, 27 galactic clusters, 6 diffuse nebulae, 4 planetary nebulae, 1 supernova remnant, 1 double star, 1 asterism, 1 bright patch of Milky Way, and at least 1 duplication. The distribution of these objects over the sky is far from uniform: clusters and diffuse nebulae are bunched along the summer Milky Way, while galaxies abound in the region of Virgo and Coma Berenices. Sagittarius alone contains 15 Messier objects.

The Messier catalogue spans a declination range from 70° north to 35° south. At least a score of objects can be seen at any time of year, but selecting a particular type of object is not always possible. For evening observing most of the galaxies are best placed in the spring, the globular clusters in the summer, and the galactic clusters in the fall and winter skies of the Northern Hemisphere.

Most of the galaxies are rather inconspicuous, and a dark night is necessary for success. Many telescope owners have located the Androm-eda spiral, M31, and have picked out the associated galaxies in the same field, M32 and NGC 205 (sometimes called M110). Fewer have found the widely illustrated M33 in nearby Triangulum. A perfect night and low power are required for this giant. Like the other spirals, it appears as a faint patch of light without detail, but much larger than most because it is one of the nearest, 2,300,000 light-years away.

Those seeking an unusual spiral system should be able to locate M51 without much difficulty; it is the Whirlpool nebula in Canes Venatici, whose irregular companion *appears* to be whirling out from the main spiral. In a 6-inch telescope, both M51 and its irregular companion are readily discernible. M64 will please those looking for a brighter galaxy after viewing the faint specks of the main Coma-Virgo hunting ground. This galaxy, sometimes called the Black-eye nebula, can be found near the star 35 Comae Berenices. The two northernmost Messier objects, M81 and M82, are galaxies that will usually impress visitors. The first is a

194

spiral, while M82 is a bright irregular system. With a rather low power, both may be maneuvered into the same field.

As summer approaches, northern observers replace low powers with high in order to resolve some of the globular clusters Messier charted. Most amateurs have examined M13, the great cluster in Hercules, which may be seen with the naked eye. Many are not aware that M4 in Scorpius, M22 in Sagittarius, and some of the Ophiuchus globulars are actually easier to resolve. Two forerunners of the main group of summer globulars give pleasant variety to spring observing: M3 in Canes Venatici, and M53 in Coma Berenices.

Messier's open clusters are concentrated along the Milky Way and are therefore almost absent in the spring when it lies along the horizon. Some Messier clusters resemble random patches of Milky Way so much that identification becomes confusing, but M6 and M7, though deep in the Scorpius star-clouds, are striking enough for easy recognition. A fine field of stars such as these or M44, the Praesepe in Cancer, seldom fails to be impressive with low powers.

Much variety is found among the clusters. A triplet in Auriga illustrates this: M36 and M38 are coarse clusters, while M37 appears as a fine pepper-shaker grouping. M35, which lies in Gemini nearby, fits midway in the scale. In Scutum, M11 is a fine fan-shaped selection which shows to best advantage under somewhat higher powers.

The diffuse and planetary nebulae are perhaps the most interesting of all, partly because they are rare, but mostly because an amateur telescope will show some detail. Traces of rifts and arms become visible in the Trifid nebula, M20, in a 6- or 8-inch instrument, although the division into three parts does not show. One of the few Messier objects associated with another, M20 joins the delicate open cluster M21; Messier recorded both as clusters, although he also noticed the nebulosity.

Not too much imagination is needed to discover the shape of the Omega or Horseshoe nebula, M17, in Sagittarius. Near the foot of the Northern Cross is M27, the Dumbbell nebula; the notches in this round patch should be evident in a good telescope. This planetary is larger and brighter than M57, the famed Ring nebula in nearby Lyra, yet it is not observed nearly as often. In Taurus, M1, the Crab nebula, is for the hale and hearty who brave the winter weather. Its pseudoplanetary appearance makes it an interesting object. In the same region of the sky, M42 and M43, both numbers referring to parts of the Orion nebula, are usually well observed.

On the following two pages is a simplified constellation chart on which all the Messier objects have been plotted. And beginning on page 198 is a checklist that will allow observers to record the objects found and the circumstances under which they were seen.

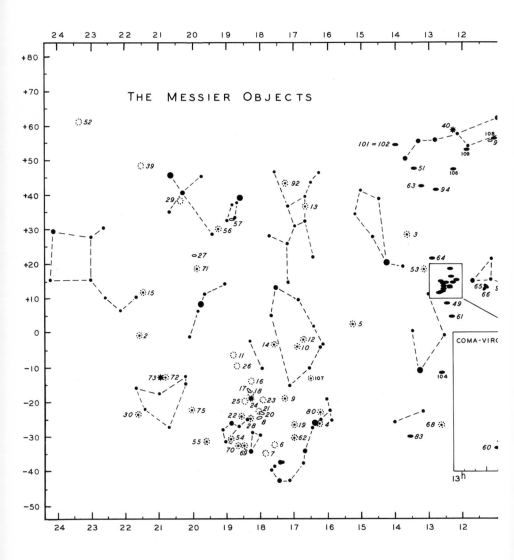

THE MESSIER OBJECTS

196

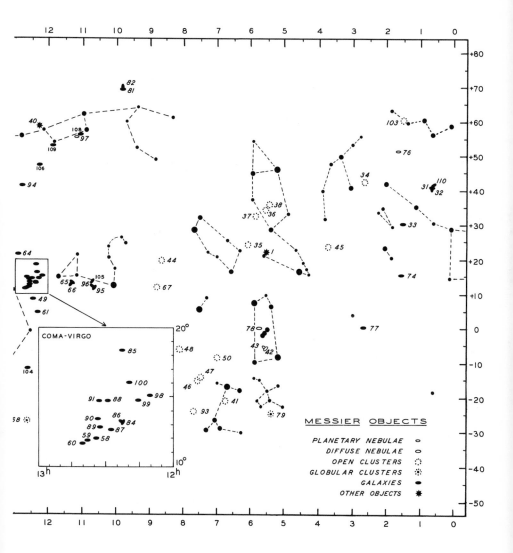

MESSIER OBJECTS

PLANETARY NEBULAE	○
DIFFUSE NEBULAE	○
OPEN CLUSTERS	⦂
GLOBULAR CLUSTERS	⦂
GALAXIES	⬤
OTHER OBJECTS	✳

COMA-VIRGO

A Messier Checklist

	Date Observed	Telescope and Magnification	Seeing	Trans.
M1				
M2				
M3				
M4				
M5				
M6				
M7				
M8				
M9				
M10				
M11				
M12				
M13				
M14				
M15				
M16				
M17				

Notes on My Observations

M1 _____

M2 _____

M3 _____

M4 _____

M5 _____

M6 _____

M7 _____

M8 _____

M9 _____

M10 _____

M11 _____

M12 _____

M13 _____

M14 _____

M15 _____

M16 _____

M17 _____

	Date Observed	Telescope and Magnification	Seeing	Trans.
M18				
M19				
M20				
M21				
M22				
M23				
M24				
M25				
M26				
M27				
M28				
M29				
M30				
M31				
M32				
M33				
M34				
M35				
M36				
M37				
M38				
M39				
M40				
M41				

Notes on My Observations

M18 _____

M19 _____

M20 _____

M21 _____

M22 _____

M23 _____

M24 _____

M25 _____

M26 _____

M27 _____

M28 _____

M29 _____

M30 _____

M31 _____

M32 _____

M33 _____

M34 _____

M35 _____

M36 _____

M37 _____

M38 _____

M39 _____

M40 _____

M41 _____

	Date Observed	Telescope and Magnification	Seeing	Trans.
M42				
M43				
M44				
M45				
M46				
M47				
M48				
M49				
M50				
M51				
M52				
M53				
M54				
M55				
M56				
M57				
M58				
M59				
M60				
M61				
M62				
M63				
M64				
M65				

Notes on My Observations

M42 _____

M43 _____

M44 _____

M45 _____

M46 _____

M47 _____

M48 _____

M49 _____

M50 _____

M51 _____

M52 _____

M53 _____

M54 _____

M55 _____

M56 _____

M57 _____

M58 _____

M59 _____

M60 _____

M61 _____

M62 _____

M63 _____

M64 _____

M65 _____

	Date Observed	Telescope and Magnification	Seeing	Trans.
M66				
M67				
M68				
M69				
M70				
M71				
M72				
M73				
M74				
M75				
M76				
M77				
M78				
M79				
M80				
M81				
M82				
M83				
M84				
M85				
M86				
M87				
M88				
M89				

M66 _____

M67 _____

M68 _____

M69 _____

M70 _____

M71 _____

M72 _____

M73 _____

M74 _____

M75 _____

M76 _____

M77 _____

M78 _____

M79 _____

M80 _____

M81 _____

M82 _____

M83 _____

M84 _____

M85 _____

M86 _____

M87 _____

M88 _____

M89 _____

	Date Observed	Telescope and Magnification	Seeing	Trans.
M90	_____	_____	_____	_____
M91	_____	_____	_____	_____
M92	_____	_____	_____	_____
M93	_____	_____	_____	_____
M94	_____	_____	_____	_____
M95	_____	_____	_____	_____
M96	_____	_____	_____	_____
M97	_____	_____	_____	_____
M98	_____	_____	_____	_____
M99	_____	_____	_____	_____
M100	_____	_____	_____	_____
M101	_____	_____	_____	_____
M102				
M103	_____	_____	_____	_____
M104	_____	_____	_____	_____
M105	_____	_____	_____	_____
M106	_____	_____	_____	_____
M107	_____	_____	_____	_____
M108	_____	_____	_____	_____
M109	_____	_____	_____	_____
M110	_____	_____	_____	_____

Notes on My Observations

M90 _____

M91 _____

M92 _____

M93 _____

M94 _____

M95 _____

M96 _____

M97 _____

M98 _____

M99 _____

M100 _____

M101 _____

M102

M103 _____

M104 _____

M105 _____

M106 _____

M107 _____

M108 _____

M109 _____

M110 _____

Additional Observing Notes

Additional Observing Notes

_____

_____

_____

_____

_____

_____

_____

_____

_____

_____

_____

_____

_____

_____

_____

_____

_____

_____

_____

_____

_____

_____

_____

_____

_____

Index and Technical Data
About the Album Photographs

TECHNICAL DATA ABOUT THE ALBUM PHOTOGRAPHS

	NGC	Coordinates (2000) R.A. h	m	Dec. °	'	Photographic Information Exposure[1] Date	Min.	Development Min.	Developer	Reproduction Scale (arc min./in.)
M1	1952	5	34.5	+22	01	Dec. —, 1965	10	2	D-19	2.8
M2	7089	21	33.5	— 0	49	June 25, 1966	10	10	D-76 (1:4)[2]	3.5
M3	5272	13	42.2	+28	23	Apr. 25, 1966	20	10	D-76 (1:4)	4.4
M4	6121	16	23.6	—26	31	July 6, 1966	5	2	D-19	4.3
M5	5904	15	18.5	+ 2	05	May 14, 1966	10	10	D-76 (1:4)	4.3
M6	6405	17	40.0	—32	12	May 14, 1966	10	2	D-19	7.9
M7	6475	17	54.0	—34	49	May 14, 1966	10	2	D-19	7.7
M8	6523	18	03.7	—24	23	July 7, 1964	40	4	D-19	8.2
M9	6333	17	19.2	—18	31	May 19, 1966	20	10	D-76 (1:4)	4.4
M10	6254	16	57.2	— 4	06	Apr. 28, 1966	20	10	D-76 (1:4)	3.8
M11	6705	18	51.1	— 6	16	June 23, 1966	5	4	D-19	8.6
M12	6218	16	47.2	— 1	57	Apr. 28, 1966	20	10	D-76 (1:4)	4.5
M13	6205	16	41.7	+36	28	Apr. 30, 1966	20	3½	D-76	4.0
M14	6402	17	37.6	— 3	15	June 18, 1966	20	10	D-76 (1:4)	4.4
M15	7078	21	30.0	+12	10	Sept. 10, 1964	30	2½	UFG	3.6
M16	6611	18	18.9	—13	47	May 28, 1966	10	4	D-19	8.9
M17	6618	18	20.8	—16	10	May 28, 1966	10	4	D-19[3]	8.6
M18	6613	18	19.9	—17	08	June 21, 1966	10	4	D-19	8.7
M19	6273	17	02.6	—26	16	June 10, 1966	20	10	D-76 (1:4)	4.5
M20	6514	18	02.4	—23	02	May 28, 1966	10	4	D-19[3]	6.8
M21	6531	18	04.7	—22	30	June 20, 1966	5	4	D-19	9.0
M22	6656	18	36.4	—23	54	May 29, 1966	10	10	D-76 (1:4)	3.5
M23	6494	17	56.9	—19	01	June 20, 1966	5	4	D-19	8.6
M24	6603	18	18.4	—18	25	May 29, 1966	5	4	D-19	7.4
M25	IC 4725	18	31.7	—19	14	June 23, 1966	5	4	D-19	8.7
M26	6694	18	45.2	— 9	24	June 23, 1966	5	4	D-19	8.8
M27	6853	19	59.6	—22	43	June 12, 1964	40	3	D-19	3.4
M28	6626	18	24.6	—24	52	June 21, 1966	10	2	D-19	3.6
M29	6913	20	24.0	+38	31	June 26, 1966	10	4	D-19	8.5
M30	7099	21	40.4	—23	11	July 14, 1966	10	3½	D-76	4.4

TECHNICAL DATA ABOUT THE ALBUM PHOTOGRAPHS

	NGC	Coordinates (2000) R.A. h	m	Dec. °	'	Photographic Information Exposure[1] Date	Min.	Development Min.	Developer	Reproduction Scale (arc min./in.)
131	224	0	42.7	+41	16	Dec. —, 1963	60	4	D-19[4]	10.7
132	221	0	42.7	+40	52	Dec. 3, 1963	40	4	D-19	3.4
133	598	1	33.8	+30	39	Dec. 6, 1963	60	4	D-19	8.4
134	1039	2	42.0	+42	47	Aug. 25, 1966	5	4	D-19	11.0
135	2168	6	08.8	+24	20	Dec. 24, 1963	15	4	D-19	8.5
136	1960	5	36.3	+34	08	Oct. 20, 1966	5	4	D-19	10.7
137	2099	5	53.0	+32	33	Oct. 20, 1966	5	4	D-19	8.6
138	1912	5	28.7	+35	50	Oct. 21, 1966	5	4	D-19	8.5
139	7092	21	32.3	+48	26	June 28, 1966	10	2	D-19	8.8
140	WNC 4	12	22.2	+58	05	June 13, 1966	10	2	D-19	7.7
141	2287	6	47.0	−20	46	Oct. 26, 1966	5	4	D-19	8.6
142	1976	5	35.3	− 5	23	Oct. —, 1965	10	10	D-76 (1:5)	6.5
143	1982	5	35.5	− 5	16	Oct. —, 1965	10	10	D-76 (1:5)	6.5
144	2632	8	40	+20.0		Apr. 13, 1966	10	4	D-19	10.9
145	η Tau	3	47.5	+24	07	Dec. 3, 1965	10	4	D-19[3]	8.9
146	2437	7	41.8	−14	49	—	10	4	D-19	8.6
147	2422	7	36.6	−14	29	—	—	—	—	8.6
148	2548	8	13.8	− 5	48	Apr. 13, 1966	5	4	D-19	11.0
149	4472	12	29.8	+ 8	00	—	20	10	D-76 (1:4)[5]	3.5
150	2323	7	03.0	− 8	21	Oct. 23, 1966	5	4	D-19	8.8
151	5194	13	29.9	+47	12	May 27, 1966	10	2	D-19	3.4
152	7654	23	24.2	+61	36	June 28, 1966	10	2	D-19	8.8
153	5024	13	12.9	+18	10	May 16, 1966	20	10	D-76 (1:5)	4.5
154	6715	18	55.1	−30	28	June 23, 1966	10	10	D-76 (1:4)	4.5
155	6809	19	40.0	−30	57	June 25, 1966	10	10	D-76 (1:4)	4.3
156	6779	19	16.6	+30	11	June 26, 1966	10	3½	D-76	4.3
157	6720	18	53.6	+33	02	June 21, 1964	15	3	UFG	2.0
158	4579	12	37.7	+11	49	June 20, 1966	12	2	D-19	3.4
159	4621	12	42.0	+11	39	Mar. 21, 1966	10	2	D-19	5.9
160	4649	12	43.7	+11	33	Mar. 21, 1966	10	2	D-19	4.2

TECHNICAL DATA ABOUT THE ALBUM PHOTOGRAPHS

	NGC	Coordinates (2000) R.A. h	m	Dec. °	,	Photographic Information Exposure[1] Date	Min.	Development Min.	Developer	Reproduction Scale (arc min./in.)
M61	4303	12	21.9	+ 4	28	June 10, 1966	20	10	D-76 (1:4)	2.7
M62	6266	17	01.2	−30	07	May 13, 1966	10	3½	D-76	5.0
M63	5055	13	15.8	+42	02	Mar. 21, 1966	10	2	D-19	3.5
M64	4826	12	56.7	+21	41	Apr. 15, 1966	20	10	D-76 (1:4)	2.9
M65	3623	11	18.9	+13	06	Feb. 15, 1964	40	—	UFG	3.8
M66	3627	11	20.3	+13	00	Feb. 15, 1964	40	—	UFG	3.7
M67	2682	8	51.3	+11	48	Apr. 13, 1966	10	4	D-19	9.8
M68	4590	12	39.5	−26	45	—	—	—	—	4.2
M69	6637	18	31.4	−32	21	June 21, 1966	5	2	D-19	4.3
M70	6681	18	43.2	−32	17	June 23, 1966	10	10	D-76 (1:4)	3.5
M71	6838	19	53.7	+18	47	June 26, 1966	10	3½	D-76	4.4
M72	6981	20	53.5	−12	32	June 25, 1966	10	3½	D-76	5.4
M73	6994	20	59.0	−12	38	June 25, 1966	5	3½	D-76	8.5
M74	628	1	36.7	+15	47	Oct. 19, 1966	20	10	D-76 (1:2)	3.4
M75	6864	20	06.1	−21	55	June 25, 1966	10	3½	D-76	3.5
M76	650-1	1	42.2	+51	34	Oct. 4, 1964	40	3	D-19	2.4
M77	1068	2	42.7	− 0	01	Oct. 19, 1966	20	10	D-76 (1:2)	3.9
M78	2068	5	46.7	+ 0	04	Oct. 22, 1966	15	4	D-19[3]	10.4
M79	1904	5	24.2	−24	31	—	—	—	—	4.2
M80	6093	16	17.0	−22	59	May 13, 1966	10	3½	D-76	4.4
M81	3031	9	55.8	+69	04	Dec. 5, 1965	20	2	D-19	5.6
M82	3034	9	56.2	+69	42	Dec. 4, 1965	20	10	D-76 (1:4)	3.2
M83	5236	13	37.7	−29	52	—	—	—	—[6]	3.4
M84	4374	12	25.1	+12	53	Mar. 21, 1966	12	2	D-19	8.6
M85	4382	12	25.4	+18	11	—	—	—	—[7]	3.5
M86	4406	12	26.2	+12	57	Mar. 21, 1966	12	2	D-19	4.4
M87	4486	12	30.8	+12	23	June 7, 1966	20	10	D-76 (1:4)	2.4
M88	4501	12	32.0	+14	25	May 17, 1966	20	10	D-76 (1:4)	3.7
M89	4552	12	35.7	+12	33	June 20, 1966	12	2	D-19	3.5
M90	4569	12	36.8	+13	10	May 19, 1966	20	10	D-76 (1:4)	3.2

TECHNICAL DATA ABOUT THE ALBUM PHOTOGRAPHS

	NGC	R.A. h	m	Dec. °	'	Exposure Date	Min.	Min.	Developer	Reproduction Scale (arc min./in.)
M91	4548	12	35.4	+14	30	Mar. 14, 1970	20^8	4	D-19^8	2.2
M92	6341	17	17.1	+43	08	June 18, 1966	20	10	D-76 (1:4)	4.4
M93	2447	7	44.6	−23	53	Oct. 23, 1966	5	4	D-19	8.8
M94	4736	12	50.9	+41	07	Feb. 22, 1964	60	3½	UFG	2.8
M95	3351	10	44.0	+11	42	—^{9}	20	10	D-76 (1:4)	3.5
M96	3368	10	46.8	+11	49	June 9, 1966	20	10	D-76 (1:4)	2.8
M97	3587	11	14.9	+55	01	Dec. 5, 1965	20	2	D-19	2.8
M98	4192	12	13.8	+14	54	Apr. 14, 1966	20	10	D-76 (1:4)	3.2
M99	4254	12	18.8	+14	25	Apr. 14, 1966	20	10	D-76 (1:4)	3.4
M100	4321	12	22.9	+15	49	May 16, 1966	20	10	D-76 (1:4)	3.4
M101	5457	14	03.2	+54	21	May —, 1965	10	—	D-19	7.3
M102	—									
M103	581	1	33.1	+60	42	July 15, 1966	5	4	D-19	8.7
M104	4594	12	40.0	−11	37	Apr. —, 1964	—	—	D-19	3.6
M105	3379	10	47.9	+12	35	Apr. 12, 1966	20	10	D-76 (1:4)	3.4
M106	4258	12	19.0	+47	18	Feb. 22, 1964	60	3½	UFG	5.8
M107	6171	16	32.5	−13	03	May 13, 1966	20	3½	D-76	4.1
M108	3556	11	11.6	+55	40	Apr. 23, 1966	20	10	D-76 (1:4)	2.6
M109	3992	11	57.7	+53	22	May 13, 1966	20	3½	D-76	2.8
M110	205	0	40.3	+41	41	Jan. 1, 1970	15^8	4	D-19	7.0

Coordinates (2000) — Photographic Information — Exposure[1] — Development — Reproduction Scale

. The Kreimer cooled camera was used for the exposures in dark type.

. Developer to water dilution ratio.

. Dodging mask used in making the print.

. Composite print from three negatives.

. Two negatives, taken May 18 and June 11, 1966, were superimposed to increase contrast.

. Two negatives were superimposed to increase contrast: April 25, 1966, a 20-minute exposure with the cooled camera, developed in D-76 (1:4) for 10 minutes, and June 9, 1966, a 10-minute exposure with the cooled camera, developed in D-19 for two minutes.

. Two negatives were superimposed: May 17, 1966, a 20-minute exposure, developed in D-76 (1:4) for 10 minutes, and June 11, 1966, a 20-minute exposure developed identically.

. Taken with the Richard Lines 16-inch f/8 reflector.

. Two negatives superimposed to increase contrast, taken June 10, 1966, and June 11, 1966.

M1   EVERED KREIMER

BEN MAYER

M4

GEORGE EAST

MILKY
WAY

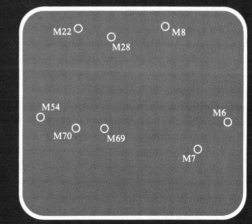

M22 ○   ○ M28   ○ M8

M54 ○   ○ M6
○ M70   ○ M69
○ M7

DAVID HEALY

M7

BRAD WALLIS, ROBERT PROVIN

BEN MAYER

M8

M8

M20 M21

ASTROPHOTO LABORATORY

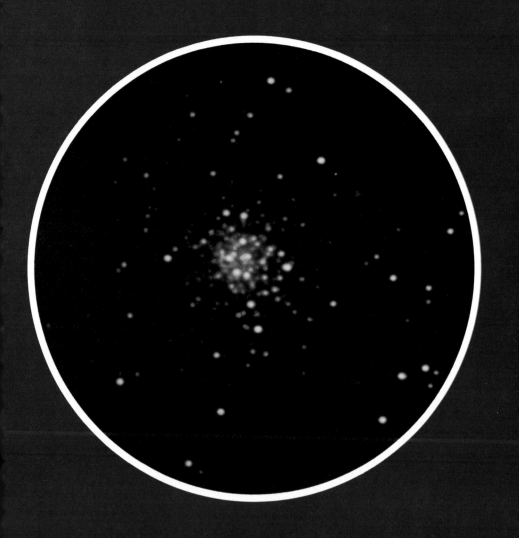

BEN MAYER

M10

M13

BEN MAYER

MILKY
WAY

DAVID HEALY

M16
M17 M18
M24
M25
M23
M21 M20
M22
M8
M28

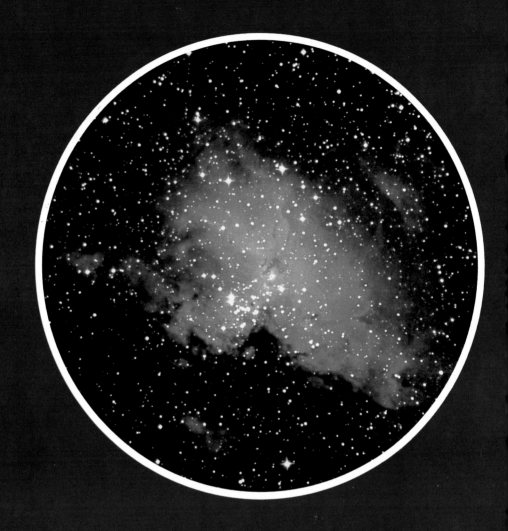

M 16   ASTROPHOTO LABORATORY

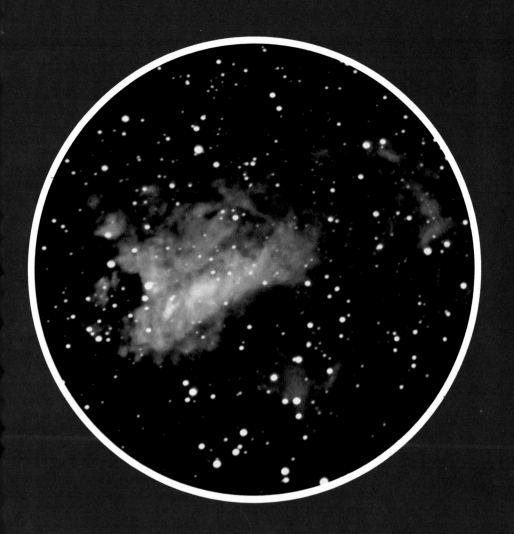

BEN MAYER

M17

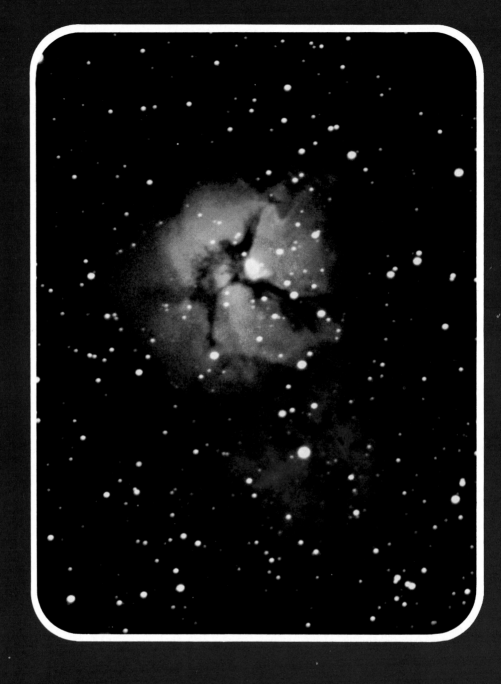

M20

BEN MAYER

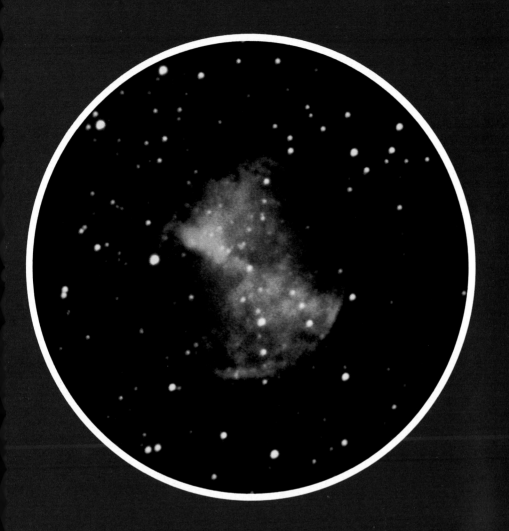

BEN MAYER

M27

M32

M110 M31

HANS VEHRENBERG

EVERED KREIMER

M33

M35

BRAD WALLIS, ROBERT PROVIN

M37

M38

BRAD WALLIS, ROBERT PROVIN

EVERED KREIMER

M42

M43

M45

ASTROPHOTO LABORATORY

EVERED KREIMER

M46

M51

BEN MAYER

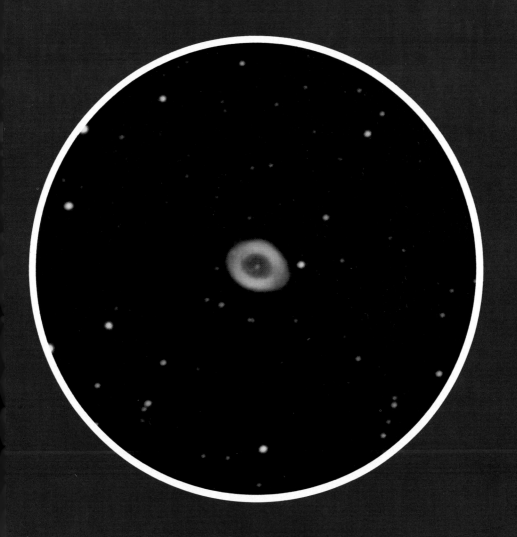

BEN MAYER

M57

M62

BEN MAYER

EVERED KREIMER

M65 M66
NGC 3628

M80

BEN MAYER

EVERED KREIMER

M81

M82

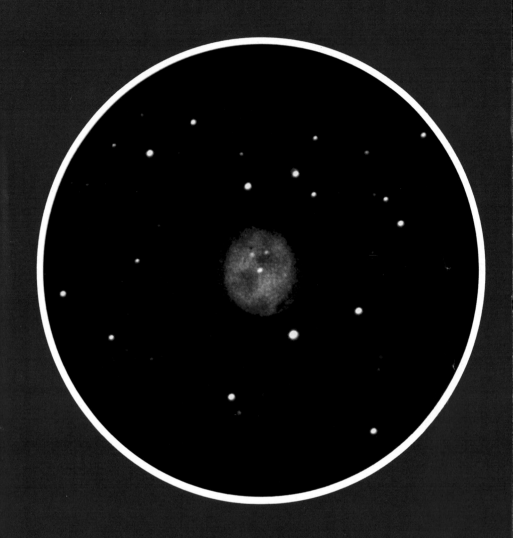

M97

EVERED KREIMER

EVERED KREIMER

M101

M104

BEN MAYER